Gas Metal Arc Welding Handbook

William H. Minnick

Professor Emeritus
Palomar College
San Marcos, California

Publisher
The Goodheart-Willcox Company, Inc.
Tinley Park, Illinois
www.g-w.com

The Goodheart-Willcox Company, Inc. Brand Disclaimer: Brand names, company names, and illustrations for products and services included in this text are provided for educational purposes only, and do not represent or imply endorsement or recommendation by the author or the publisher.

The Goodheart-Willcox Company, Inc. Safety Notice: The reader is expressly advised to carefully read, understand, and apply all safety precautions and warnings described in this book or that might also be indicated in undertaking the activities and exercises described herein to minimize risk of personal injury or injury to others. Common sense and good judgment should also be exercised and applied to help avoid all potential hazards. The reader should always refer to the appropriate manufacturer's technical information, directions, and recommendations; then proceed with care to follow specific equipment operating instructions. The reader should understand these notices and cautions are not exhaustive.

The publisher makes no warranty or representation whatsoever, either expressed or implied, including, but not limited to, equipment, procedures, and applications described or referred to herein, their quality, performance, merchantability, or fitness for a particular purpose. The publisher assumes no responsibility for any changes, errors, or omissions in this book. The publisher specifically disclaims any liability whatsoever, including any direct, indirect, incidental, consequential, special, or exemplary damages resulting, in whole or in part, from the reader's use or reliance upon the information, instructions, procedures, warnings, cautions, applications, or other matter contained in this book. The publisher assumes no responsibility for the activities of the reader.

Cover image courtesy of Lincoln Electric Co.

Library of Congress Cataloging-in-Publication Data
Minnick, William H.
    Gas metal arc welding handbook / by William H. Minnick.
— 5th ed.
      p. cm.
    Includes index.
    ISBN-13: 978-1-59070-866-8
     1. Gas metal arc welding—Handbooks, manuals, etc.
I. Title.
TK4660.M528 2007
671.5'212—dc22   2007039796

Introduction

Gas Metal Arc Welding Handbook provides a simple but complete introduction to Gas Metal Arc Welding (GMAW), which has long been an important welding process. Improvements in GMAW over the years, as well as refinements in the equipment, have established its position as a major industrial welding system. This **Handbook** covers GMAW principles, equipment, modes of operation, and safety in a straightforward, no-nonsense manner.

The organization of the chapters in this book will lead you step-by-step through the principles and practices of this important welding process. New material is introduced as it is needed. Topics covered include:

- Basic operation of each component of the system.
- Safety practices specific to working with electricity, shielding gases, and other welding hazards.
- Various types of welds and weld joints.
- Welding techniques and procedures for several different metals.
- Weld defects and how to avoid them.
- Special procedures and techniques for welding auto bodies, trucks, and off-road vehicles.

For a beginning student of welding, the book provides the background needed for success in a welding career. Advanced students will find it valuable as a "refresher" course and for its detailed coverage on how to weld various metals using the modes common to the industry.

A careful study of this **Gas Metal Arc Welding Handbook** and practice of the welding procedures it covers will provide you with the knowledge and skill you need to secure employment as a welder.

William H. Minnick

About the Author

William H. Minnick, realizing the need for a specialized type of welding text for instructors and students, has drawn upon his many years of experience as a welder, welding engineer, and community college instructor to develop this text for training future welders.

The author's career in industry includes experience in welding on jet engines, missiles, pressure vessels, and atomic reactors. He developed the welding procedures for, and welded, the first titanium pressure vessel for the Atlas missile program. He has written many technical papers, including articles on research and development for welding exotic materials and on modification of existing welding processes for automatic and robotic applications.

Mr. Minnick has developed welding certificate and degree programs, and taught all phases of welding and metallurgy in community colleges for more than twenty years. In addition to this **Gas Metal Arc Welding Handbook**, he is also the author of the **Gas Tungsten Arc Welding Handbook** and the **Flux Cored Arc Welding Handbook.**

Contents

Acknowledgments

The author gratefully acknowledges the assistance of the following companies that contributed suggestions, ideas, photographs, and information to this textbook.

Arcal Chemical, Inc.	MK Products, Inc.
Bernard Welding Equipment Co.	Nederman, Inc.
CK Worldwide, Inc.	ODL, Inc.
CRC-Evans Co.	Smith's Heating Welding Equipment Co.
Distribution Designs, Inc.	Techalloy Maryland, Inc.
ESAB Welding & Cutting Products	Tescom Corp.
Ferro-Slick/Lube-Matic	Thermco Instrument Co.
FluidTech Corp.	Tweco Products, Inc.
G.A.L. Gage Co.	Veriflo Corporation
G.S. Parsons Co.	Victor Equipment Co.
HTP America, Inc.	Weld World Co.
Lincoln Electric Co.	Weldrite Mfg. Co.
Magnaflux Corp.	Western Enterprises Co.
Miller Electric Mfg. Co.	York Mfg. Co.

The author is deeply grateful for many tables, figures, and information from the American Welding Society, the computer drawings from my son Steven A. Minnick, and the photographic printing by my son William R. Minnick.

William H. Minnick

Gas Metal Arc Welding Process

Objectives

After studying this chapter, you will be able to:
- Define GMAW.
- List advantages and disadvantages of GMAW.
- Explain the difference between semi-automatic and automatic welding.
- Name applications where GMAW is used in industry.

Definition

As defined by the American Welding Society, **GMAW (Gas Metal Arc Welding)** is an electric arc welding process that fuses (melts) metallic parts by heating them with an arc between a continuously fed filler electrode and the work. The electrode is small in diameter, and is consumed (used up) during the process. A shielding gas protects the electrode and the molten weld metal from contamination by the surrounding atmosphere. Figure 1-1 illustrates a basic GMAW operation.

Process Names

The original development of the process led to patents by AIRCO (Air Reduction Co.). The AIRCO process was referred to as Metal Inert Gas welding or MIG welding. Many manufacturers have developed their own variations on the process, and have given the process various names. These names include the following:
- Aircomatic by the Air Reduction Company.
- Microwire by the Hobart Co.
- Millermatic by the Miller Electric Mfg. Co.
- Dipmatic by the Miller Electric Mfg.Co.
- Buried Arc CO2 by the Miller Electric Mfg. Co.

During the developmental years of the process, only inert gases (argon and helium) were used. The process

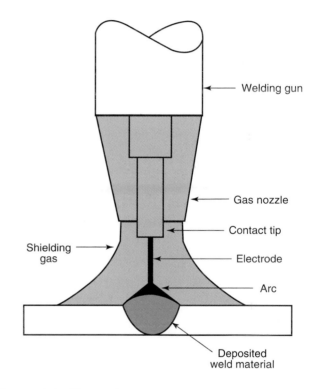

Figure 1-1. The welding gun contains components that conduct electrical power and shield the molten metal from the atmosphere.

designation, Metal Inert Gas welding, identified the gases as inert (chemically inactive). Today gases such as carbon dioxide and oxygen, which are not inert, are being used for the process. For this reason, the identification of the process has been changed to GMAW (Gas Metal Arc Welding).

The American Welding Society has further modified the GMAW identification to include additional information to identify the process type. For example, GMAW-S is used to identify the short circuiting arc mode process, and GMAW-P to identify the pulse spray arc mode process.

Modes of Operation

Semi-automatic (manual) operation is accomplished when the weld is done by a welder using the process equipment and a portable gun. An example of a semi-automatic welding operation is shown in Figure 1-2.

Automatic operation is accomplished by an operator using specialized equipment and controls that are monitored by the operator during the welding procedure. In some production welding, this specialized equipment includes computer-controlled machines or robots. An automatic machine is shown in Figure 1-3.

GMAW Equipment

The basic GMAW equipment is shown is Figure 1-4. This equipment is used for either semi-automatic or automatic welding.

Power supplies provide the type and amount of welding current to melt the base material and the filler material. The GMAW process usually uses DCRP (Direct Current Reverse Polarity), or DCEP (Direct Current Electrode Positive). The power supply shown in Figure 1-5 provides the proper type of electrical current to melt the filler material and sustain an electrical arc. The power supply is termed a *CV (constant voltage)* supply. Sometimes, it is referred to as *CP (constant potential)*.

The *wire feeder* is an electrical mechanical device that feeds the required amount of filler material at a constant rate of speed throughout the welding operation.

Figure 1-3. Automatic welding reduces welding time and increases the quality of the weld. (Jetline Engineering, Inc.)

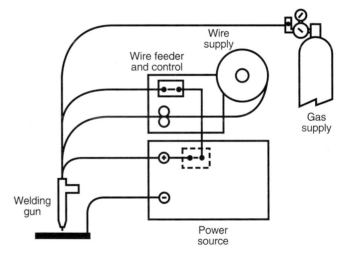

Figure 1-4. The GMAW welding equipment controls the power supply, the wire feed, and shielding gas supply.

The wire feeder shown in Figure 1-6 is typical of those found in the welding industry. The filler material is supplied in coils or wound on spools that are mounted on the wire feeder.

The *welding filler electrode* is a small diameter consumable electrode that is supplied to the welding gun by the drive roller system within the wire feeder. The electrode is usually chemically similar or identical to the work (base metal).

Figure 1-2. Semi-automatic welding is performed by a welder using a handheld welding gun.

Figure 1-5. This GMAW welding power supply can be used for either semi-automatic or automatic welding operations. (Miller Electric Mfg. Co.)

Figure 1-6. A basic wire feeder supplies the wire filler electrode at a constant rate of speed. (Miller Electric Mfg. Co.)

Shielding gas is a gas used to protect the weld (molten metal) from atmospheric contamination. It protects the wire and the weld metal from oxidation. *Oxidation* is a reaction caused by the introduction of oxygen to a substance. In welding, oxidation can adversely affect the strength of a weld. A supply of shielding gas and the regulation equipment is required to admit a continuous flow of gas during the entire welding operation. Figure 1-7 shows a supply cylinder of gas and the regulation equipment mounted on a cart with a wire feeder and power supply.

A *gun* is a trigger- or switch-operated mechanism, held by an operator or mounted on a machine, that transfers electrical current to the wire. A gas nozzle attached to the end directs the shielding gas around the weld. Figure 1-8 shows a typical welding gun used by a welder during the welding operation.

Systems

The welding equipment components are combined into systems to operate at the proper time to make a satisfactory weld. Many systems are designed for particular welding modes and uses for efficient use of the equipment.

A typical semi-automatic system is shown in Figure 1-9. It is used for welding all types of metals and varying thicknesses.

Figure 1-7. This cylinder of shielding gas with a regulator attached is mounted on a cart with power supply and wire feeder. (Lincoln Electric Co.)

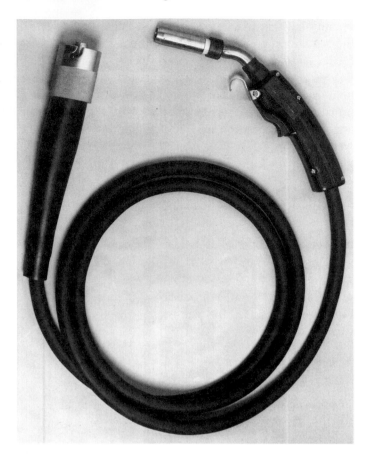

Figure 1-8. Basic welding gun with connections for power, wire supply, and shielding gas. (Bernard Welding Equipment Co.)

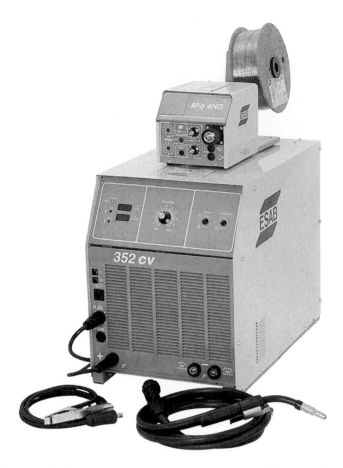

Figure 1-9. This system can use small or large diameter wire electrodes for thin or heavy material and different welding modes. (ESAB Welding & Cutting Products)

Figure 1-10 shows a 110V ac system for welding thin-gauge materials. These machines are well suited for the home repair, when making minor repairs to toys, bikes, etc.

Figure 1-11 shows a typical semi-automatic system. This type of system is used for welding thin-gauge metals used in unibody automobile repair.

Figure 1-12 illustrates how a manipulator can be used in conjunction with the GMAW process. The entire operation is preprogrammed so it can be completed automatically.

Figure 1-13 shows a typical application of GMAW used to fabricate a steel circular staircase framework. **It also illustrates the use of proper safety gear, which includes the leather gloves and welding shield. Safety gear should always be worn when performing a welding process.**

Advantages and Disadvantages

Over the years, use of GMAW has grown greatly. Today, because of many developments and applications, it has become one of the major welding processes used in the joining of metals. The process has some advantages and disadvantages when compared to other welding processes.

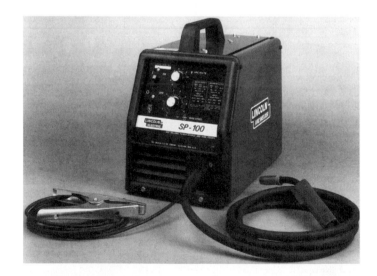

Figure 1-10. This system is a typical 110V ac welding system used for welding light gauge metals. (Lincoln Electric Co.)

Advantages

- Useful for joining major commercial metals including: carbon and alloy steels, stainless steels, aluminum, magnesium, copper, bronze, and titanium.
- Produces high-quality welds.
- In some cases, usable in all positions.

Figure 1-11. Controls have been added to this machine for spot, plug, and seam welding of thin materials. (HTP America, Inc.)

Figure 1-12. This robotic system produces consistent welds in exact programmed locations. (Miller Electric Mfg. Co.)

- Produces no slag during welding.
- High metal deposition rates.
- Delivers about 95 percent deposition efficiency.
- In some cases, may be used to bridge gaps.
- Welding techniques easily learned.

Figure 1-13. Welding a steel framework with the GMAW semi-automatic process. (Weldrite Mfg. Co.)

Disadvantages

- Requires more equipment than SMAW (Shielded Metal Arc Welding).
- Is not as portable as other welding processes.
- Requires a higher initial equipment investment.
- Subject to wind drafts that reduce the shielding of the molten metal and cause contamination of the weld.
- Greater operating expense due to high-cost shielding gases.
- Welds are prone to cold starts.
- Cannot separate wire feed speed and current level.
- Smaller diameter electrode wires are costly.

Applications

There are many applications of the GMAW process in industry, Figure 1-14. It may be used as a primary process or used in conjunction with other processes to

Figure 1-14. This welder is fabricating an assembly using the GMAW semi-automatic process. Note that the welder is properly dressed for safety, wearing eye and face protection, gloves, and leather sleeves.

weld a variety of metals. Some of the many applications include:

- Sheet metal assemblies.
- Structural steel assemblies.
- Pipelines.
- Automobiles and trucks.
- Motorcycles, off-road vehicles, and equipment.
- Campers and recreational equipment.
- Ships, barges, and boats.
- Tankers and trailers.
- Pressure vessels.
- Water tanks.
- Rockets and missiles.
- Lawn and garden equipment.

Review Questions

Write your answers on a separate sheet of paper. Do not write in this book.

1. _____ _____ _____ is the organization that defines the standard welding process.
2. The GMAW process uses a(n) _____ diameter electrode that is _____ during the process.
3. The process was originally developed by _____ and was called _____ welding.
4. Metal inert gas welding was changed to _____, because a variety of gases that are now used with the process are not inert.
5. _____ _____ and _____ are two gases that are not inert and are now used for GMAW processes.
6. Whenever a welder uses a hand-held gun, the process is termed a(n) _____-_____ welding operation.
7. A(n) _____ welding operation is accomplished by an operator using specialized equipment and controls that he or she monitors during the welding operation.
8. The term CV relates to a power supply with a(n) _____ _____ output.
9. The term CV is also referred to as CP or a(n) _____ _____ output.
10. GMAW is usually done in the DCEP, or direct current electrode _____ mode.
11. Direct current reverse polarity is also termed _____.
12. A(n) _____ _____ supplies the welding filler electrode to the welding arc.
13. Filler material is fed at a(n) _____ rate of speed by the wire feeder.
14. Shielding gas is used to prevent _____ of the molten metal during welding.
15. Shielding gas is directed to the area by the _____ _____.

2 GMAW Process Operation and Safety

Objectives

After studying this chapter, you will be able to:
- Name six major parameters and variables controlled by a welder.
- Discuss the four basic components of a GMAW system.
- Display safe welding practices in the shop.
- Eliminate unsafe conditions in the shop.

In a GMAW operation, a consumable wire electrode is fed from the wire feeder through the welding gun. Upon contact with the workpiece, heat is generated from the resistance of the electric current, which melts the electrode and the base metal to make a fusion weld. The direct current flows from the power supply through the workpiece to the electrode. Electrical current flow in this direction is called reverse polarity, or DCRP (Direct Current Reverse Polarity). Polarity means the direction in which the electricity flows. It is also referred to as DCEP (Direct Current Electrode Positive).

Metal Deposition

Various modes of metal deposition are used to deposit the weld metal into the weld joint. A weld joint is the junction of members or edges of mating parts to be joined. The metal deposition will depend on the type of metal, thickness, position of the joint, and the type of joint to be welded.

Short-Circuiting-Arc

In the *short-circuiting-arc*, or *short-arc*, mode of metal deposition, the electrode short-circuits when it is fed into the workpiece. The short circuit causes the electrode to melt through and deposit molten metal into the weld joint. Figure 2-1 illustrates this mode of metal deposition. It has many advantages, but also has disadvantages.

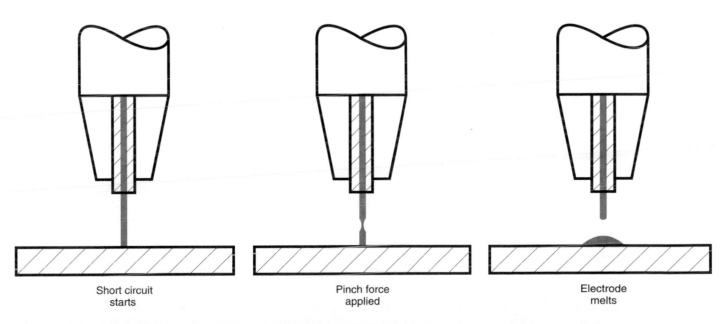

| Short circuit starts | Pinch force applied | Electrode melts |

Figure 2-1. In the short-circuiting-arc mode of metal deposition, the electrode actually touches the workpiece and melts about 20 to 200 times per second.

The advantages include:
- Low heat input.
- Welds in all positions.
- Welds thin gauge metals.
- Can be used to bridge gaps in some weld joints

The disadvantages include:
- The process uses small-diameter wire, which is expensive.
- Spatter (small globules of weld metal) may cause serious problems with various gases or techniques.
- Cold starts of the weld operation and cold laps during welding may occur due to the low heat input.
- Limited to the welding of metals that are 1/8″ thickness or less.

Spray-Arc

The *spray-arc* mode of metal deposition is a method in which the electrode is fed into the weld zone, where it is melted off above the workpiece and deposited into the weld joint. Figure 2-2 illustrates this mode of deposition. It also has many advantages and disadvantages.

The advantages include:
- Very little or no spatter.
- High-quality weld metal deposition.
- Faster welding speeds.
- Deep penetration.

The disadvantages include:
- The weld pool is very large and very fluid.
- Can only be used in flat position when welding steel groove welds.
- Can only be used in the flat and horizontal positions when making steel fillet welds.

Pulse Spray-Arc

The *pulse spray-arc* or *spray-arc pulse* mode of metal deposition is a method of pulsing a drop of filler material from the end of the electrode at a controlled time in the welding cycle. The spray-arc mode of metal deposition has been greatly improved with the addition of this method. Figure 2-3 illustrates this mode of deposition.

Its advantages include:
- Greater metal deposition can be obtained with a lower overall current level. This decreases the amount of warpage due to the lower overall amount of current needed for fusion. This warpage decrease reduces the need for welding tooling or jigging.

The disadvantages include:
- Need for special power supplies or special pulsing units that may attach to the power supply for pulsing the droplet of metal from the electrode.

Globular

The *globular* mode of metal deposition is a method in which the electrode burns off above or in contact with the workpiece in a very erratic globular pattern. Figure 2-4 illustrates this mode of deposition.

The advantages of this mode include:
- High deposition rate.
- High-quality welds are possible.

The disadvantages include:
- Considerable amount of smoke and spatter.
- Rough weld appearance.
- Can only be used in flat position.

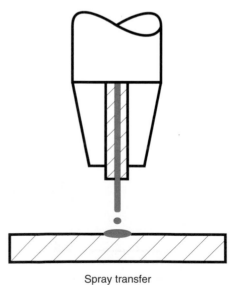

Spray transfer

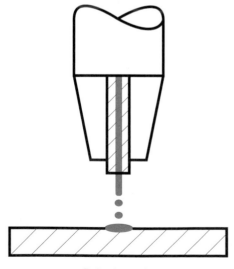

Pulsed transfer

Figure 2-2. In the spray-arc mode of metal deposition, the electrode is fed into the weld zone, melted off above the workpiece, and deposited randomly into the weld joint.

Figure 2-3. In the pulse spray-arc mode of metal deposition, drops of filler material are pulsed from the end of the electrode at a controlled time in the welding cycle.

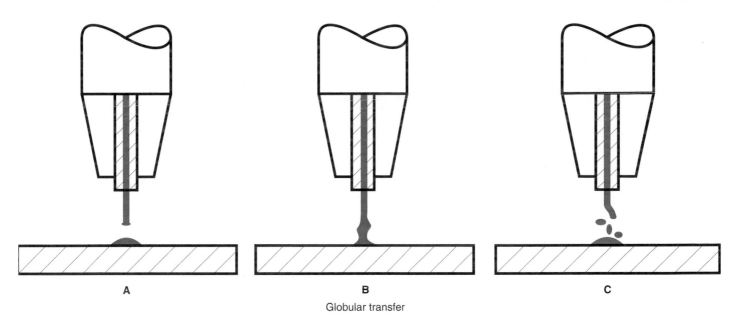

A

B

C

Globular transfer

Figure 2-4. Globular mode of deposition. A— Electrode drop forms, melts, and detaches from the high temperature of the arc. B— Electrode contacts the workpiece before melting. C— Electrode drop may disintegrate and cause spatter.

Process Setup

The proper use of any metal deposition mode or variation requires that the welding equipment be set up in a special manner for each specific application. To properly set up the main characteristics for the weld, a determination must be made of the required weld, and the machine must be set accordingly. These characteristics are called **parameters** and **variables** in the welding trade. They will vary depending on such areas as type of metal, thickness of the weld joint, type of weld joint, weld joint design, weld quality specifications, and labor cost.

The major parameters and variables include:
- Amount of weld current. (Wire speed)
- Amount of arc voltage. (Arc gap)
- Inductance. (Controls the rate of current rise in the short-arc welding cycle and controls the amount of spatter.)
- OCV and volt-ampere curve slope.
- Type of shielding gas, flow rate, and nozzle size.
- Electrode type and diameter, wire speed, contact tip size, and electrode stickout.

The reference section of this text lists the initial parameters for various types and thicknesses of metals.

Power Supplies

The basic power supply used for GMAW is designed for this particular welding process. It is called a **CP (constant potential)** or a **CV (constant voltage)** power supply. These machines supply varying amounts of amperage to maintain the preset welding voltage

(arc gap). Welding voltage is set on the machine prior to starting, and is then adjusted during the operation as required to change the arc gap dimension.

In some cases, a CC (constant current), which may also be termed a VV (variable voltage) machine, may be used as a power supply for this process. However, these machines require additional components in the system for the process to work properly.

In some cases, power supplies may include:
- **CV power supply** as shown in Figure 2-5.
- **CV power supply** combined with a wire feeder in a single unit. This is shown in Figure 2-6.
- **CC power supply** coupled with a specially designed wire feeder. This is shown in Figure 2-7.
- **CC** and **CV power supplies** are available for field welding where utility power is not available. The units are either gasoline- or diesel-powered, and may be either air- or liquid-cooled. A portable field power supply and separate wire feeder are shown in Figure 2-8.
- **Combination power supply and wire feeder** units are available for welding light gauge metals. These units operate on 110/115V ac. This type of unit is shown in Figure 2-9.
- **CV power supply** for spray-arc welding with pulse control is shown in Figure 2-10.
- **Inverter-type power supplies** are used for all types of GMAW. These machines are smaller than normal GMAW power supplies and are used where portability is an asset. A multi-welding-process power supply is shown in Figure 2-11.

Figure 2-5. A constant voltage power supply regulates amperage output automatically to maintain the arc voltage (gap) that is set on the power supply. (Miller Electric Mfg. Co.)

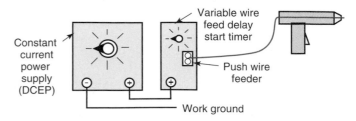

Figure 2-7. Constant current supplies can be used for GMAW, but they must have a special wire feeder for delay and regulation of the wire speed at the start of the weld operation. This type of unit can only be used to weld with the spray-arc mode.

Figure 2-8. An engine-driven power supply and separate wire feeder allow complete portability for use in the field. The wire feeder allows the operator to weld at considerable distances from the power supply. (Miller Electric Mfg. Co.)

Figure 2-6. This welding machine is actually a power supply, wire supply, and wire feeder. The unit is compact and easy to move on the job site. (Lincoln Electric)

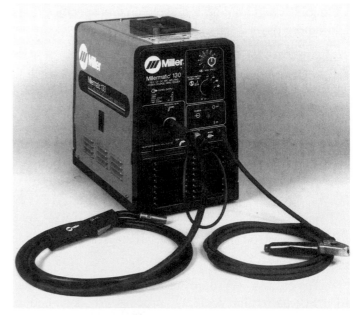

Figure 2-9. Portable-type power supplies often operate on 115V ac and are restricted to lighter-gauge metals. (Miller Electric Mfg. Co.)

Figure 2-10. This machine has been designed with special circuits for precise control of the pulsing spray-arc process. (CRC-Evans Co.)

• **Inverter-type power supplies** are also being used for pulse-spray gas metal arc welding when combined with a microprocessor wire feeder. These units are capable of being programmed at the factory for conventional GMAW and pulsed GMAW. *Synergic programs* are programs established by the factory that can be modified by the user if desired. These programs are capable of regulating changes in the welding parameters, thus obtaining the best possible welding conditions. An inverter power supply is shown in Figure 2-12A and a microprocessor wire feeder is shown in Figure 2-12B.

A

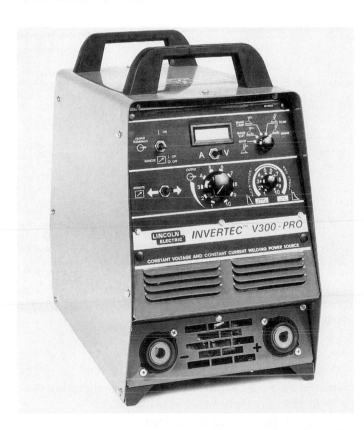

Figure 2-11. Inverter power supplies of this type have a variable inductance control for GMA short-arc welding. (Lincoln Electric)

B

Figure 2-12. A—Inverter-type power supplies can be used for conventional GMAW or pulsed-arc GMAW. B—Microprocessor wire feeders are used with the inverter-type machine shown in Figure 2-12A. (Miller Electric Mfg. Co.)

Wire Feeder

Wire feeders feed a controlled amount of filler material from the supply spool through the drive roll mechanism to the welding gun and finally to the welding arc. The controller can be separate as shown in Figure 2-13, or a part of the welding machine. A wire feeder unit can be contained in the welding machine, as shown in Figure 2-14. The drive roller mechanism is usually powered by 110V ac or 24V dc.

Gas Supply and Regulation

Shielding gases prevent contamination of the welding electrode and the weld. Argon, carbon dioxide, helium, and oxygen are the most common gases used. They are available from the manufacturer as single or mixed gases, in either liquid or gaseous form. They are supplied in pressurized cylinders of various sizes, at pressures up to 3000 psi (pounds per square inch). *Regulators* are mechanisms used to reduce the pressure of the gas to a workable level. A *flowmeter* is used to regulate the amount of gas flow to the weld area. The regulator-flowmeter shown in Figure 2-15 is commonly used with high-pressure cylinders.

To create a good weld, sufficient gas must be delivered to the welding area to prevent contamination. Too little gas flow allows the atmosphere (air) to enter the weld zone and contaminate the weld. Too much gas flow causes turbulence in the gas envelope (shield). This allows the atmosphere to enter the weld zone and also contaminates the weld. *Turbulence* is a great change in speed and direction of flow. Whenever atmospheric gases enter the shielded gas envelope, the quality of the weld decreases.

Electrically operated valves, called *solenoids*, are included in the system to start and stop the flow of shielding gas from the gas supply to the welding gun. These valves, as shown in Figure 2-16, are usually located in the wire feeder or in the wire supply cabinet. These valves are actuated by closing the switch on the welding gun or by closing a switch in the arc starting circuit.

Welding Guns

The GMAW process can use a variety of welding guns to deliver the welding wire to the arc. These units are designed and constructed to protect the user from electrical shock and to provide a means to conduct electrical current to the consumable welding wire as it passes through the *contact tip* in the end of the welding gun.

Welding guns are rated by the amperage they can carry and for the length of time they can be used at that amperage. The guns used for intermittent and low-amperage welding are cooled by the shielding gas, while others used for heavy-duty welding are water-cooled.

Figure 2-13. This wire feeder has a rheostat for adjusting the wire feed. A switch is used for inching the wire and for purging the shielding gas system.

Figure 2-14. This wire feeder has an adapter that will allow either a small or large wire spool to be used. The small spool is in place. (Miller Electric Mfg. Co.)

The guns are rated by duty cycle, which is the same manner the power supplies are rated. The gun rating may be from 20% to 100% duty cycle.

Gas nozzles of various designs attach to the welding gun to direct the shielding gas around the welding wire and the molten metal. Contact tips are attached to the gun within the nozzle area. The tip transfers welding current

Figure 2-15. Single-cylinder regulator/flowmeters have a gauge that shows cylinder pressure. Gas flow (in cubic feet per hour) to the gun is adjusted by turning the adjustment knob and reading the ball position in the vertical tube. (CONCOA)

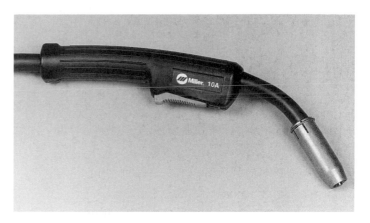

Figure 2-17. GMAW guns of this type are cooled by the passage of shielding gas through the handle and nozzle assembly. (Miller Electric Mfg. Co.)

Figure 2-16. When connecting valves to the service line, always use plastic tape on the male fitting. Shielding gas leaks may cause contamination of the weld, leading to porosity and cracks. Check often for leaks and repair as needed.

Figure 2-18. The nozzle of this gun is designed to remove the smoke and fumes from the welding operation. A special vacuum system attached to the gun draws in the fumes. Inset— Note the diagonal smoke-extracting inlets above the gas nozzle opening. (Tweco-Airarc, a Thermadyne Company)

to the wire as the wire passes through it. A typical GMAW process welding gun is shown in Figure 2-17.

Some welding operations create considerable smoke or fumes, due to the type of material, process mode, or use of specific types of shielding gases. To reduce the operator contact with these pollutants, special guns have been developed. The gun shown in Figure 2-18 has a special nozzle built into its tip. It is connected to a vacuum machine to remove the smoke and fumes developed

during welding. The vacuum system shown in Figure 2-19 may also be used for this purpose.

The welding gun shown in Figure 2-20 draws the wire from the wire spool to the set of drive rollers mounted in the gun housing. These guns usually have the capability for water-cooling for an increased duty cycle. A typical water cooler for this system is shown in Figure 2-21.

Figure 2-19. This type of system uses a large hood, rather than the special gun nozzle, to remove the smoke. (Nederman, Inc.)

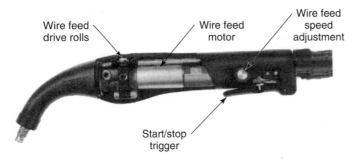

Wire feed drive rolls

Wire feed motor

Wire feed speed adjustment

Start/stop trigger

Figure 2-20. A low voltage motor and drive rolls within the gun body pull the wire from the supply spool, and push the wire to the contact tip. Water hoses are attached to the rear of the gun for cooling the nozzle body, which gives the gun a very high duty cycle. (MK Products, Inc.)

Figure 2-21. Water coolers of this type operate on 110V ac and are very efficient. They are portable and easily installed on water-cooled guns. (Tweco-Airarc, a Thermadyne Company)

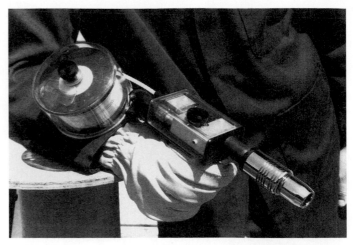

Figure 2-22. An electric motor drives the wire feeder at a speed set on the adjustable rheostat on the gun casing. (CK Worldwide, Inc.)

A *portable* gun unit containing the welding wire and the drive rolls is shown in Figure 2-22 and is termed a **spool gun**. The wire speed is controlled by an adjustable rheostat mounted on the gun.

When welding with automatic or robotic-type machines, a gun similar to the one shown in Figure 2-23 would be used. These machines operate at higher current values for longer periods of time, therefore they are usually water-cooled.

GMAW Safety

GMAW is a skill that may be performed safely with minimum risk, if the welder uses common sense and follows safety rules. It is recommended that you establish safety habits as you work with this process. Check equipment regularly and be sure that your environment is safe. GMAW safety covers four major areas and includes:

1. **Electrical Current Safety.** Primary voltage to the electrically powered welding machine is usually 115V ac to 440V ac, and in some cases may even be more. This amount of voltage may cause extreme shock to the body and possibly death. For this reason, all safety rules should be followed.
   A. Never install fuses of a higher amperage than specified on the data label or the operation manual.
   B. Always use a ground wire from the machine frame to a suitable solid ground.
   C. Install electrical components in compliance with all electrical codes, rules, and regulations.
   D. Be certain that all electrical connections are tight.
   E. Never open a welding machine cabinet when the machine is operating, unless you are a competent electronic technician and are fully trained in machine installation and operation.

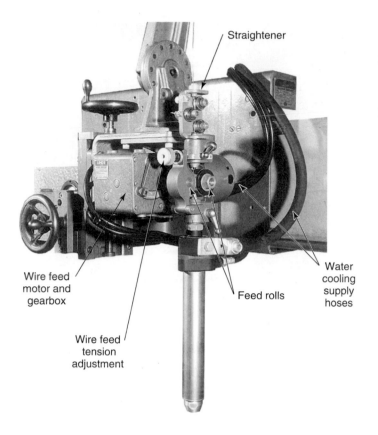

Figure 2-24. This cylinder has been made and tested to a DOT (Department of Transportation) specification. The letter "T" indicates the amount of gas that it will hold. The name imprinted is the owner.

Figure 2-23. This machine-driven gun is water-cooled to allow higher-amperage duty. The gun is mounted directly onto the wire feeder and a wire straightener is mounted on top of the feeder. A straightener is used to remove some of the curve (cast) of the wire prior to the wire entering the feed rolls.

F. Always lock primary voltage switches open and remove fuses when working on electrical components inside the welding machine. **Then put the lock key in your pocket!**

G. Welding current supplied by constant current and constant potential power supplies have an open circuit voltage range of 40 to 80 volts. At this low voltage, the possibility of lethal shock is small. However, it will still produce a good shock that may cause health problems. To reduce the possibility of this happening:

   a. Keep the welding power supply dry.

   b. Keep the power cable, the ground cable, and the gun dry.

   c. Make sure the ground clamp is securely attached to the power supply and workpiece.

2. **Shielding Gas Safety.** The gases used in GMAW are produced and distributed to the user in either liquid or gaseous form. All storage vessels used for these gases are approved by the DOT (Department of Transportation) or previously by the ICC (Interstate Commerce Commission), and are so stamped on the vessels cylinder walls. A cylinder stamping is shown in Figure 2-24. Some of the gases used in GMAW are

inert and colorless, therefore, special precautions must be taken when using them. None of the gases are toxic, but they can cause *asphyxiation* (suffocation) in a confined area without sufficient ventilation. Any atmosphere that does not contain at least 18% oxygen can cause dizziness, unconsciousness, or even death.

Shielding gases cannot be detected by the human senses and can be inhaled like air. Never enter any tank, pit, or vessel where gases may be present until the area is *purged* (cleaned) with air and checked for oxygen content. High pressure gas cylinders contain gases under very high pressure (approximately 2000 to 4000 psi) and must be handled with extreme care. Each of the following rules should be followed:

A. Store all cylinders in the vertical position.

B. Secure all cylinders with safety chains, cables, or straps. See Figure 2-25.

C. Do not use cylinders as rollers.

D. Know the contents before use. Figure 2-26 illustrates a label indicating the type of gas within the cylinder.

E. Keep the protective cap in place until the cylinder is ready to use.

F. Do not move a cylinder without the protective cap in place. Always use a cylinder cart, with the safety chains installed, to move a cylinder.

G. Check the outlet threads and clean the valve opening by *cracking* (slightly opening and closing) the cylinder valves as shown in Figure 2-27 before attaching the regulator.

H. Use only the proper equipment for the type of gas used.

I. Attach the regulator and securely mount the flowmeter in the vertical position, Figure 2-28.

Figure 2-25. Secure all cylinders with chains.

Figure 2-26. The label indicates the type of gas stored, a caution, and the cylinder contents in cubic feet. The safety cap is in place until ready to use.

Figure 2-27. "Cracking" a cylinder before attaching a regulator blows out any dirt in the valve opening.

J. When opening a cylinder valve, the regulator adjusting screw should always be free of pressure on the regulator diaphragm.

K. Stand aside and slowly open the cylinder valve. Never stand in front of the gauges when opening a cylinder.

L. When cylinders are not in use, the cylinder valve should be closed and the diaphragm screw should be fully loosened.

M. When cylinder is empty, close the valve, replace the safety cap, and mark the letters "MT" with chalk on the upper part of the cylinder.

N. Never tamper with a leaky valve; return the cylinder to the supplier for a replacement.

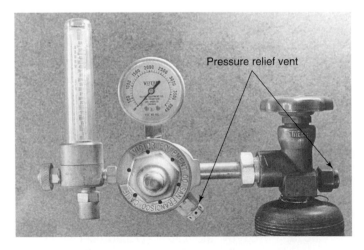

Figure 2-28. This regulator/flowmeter has been properly attached to the cylinder. Note the pressure relief vent on the regulator and cylinder. These devices operate to release excess pressure in the cylinder. Never tamper with these devices. (Victor Equipment Co.)

Liquefied gas cylinders are commonly called ***Dewar flasks*** and are basically vacuum bottles. The gas has been reduced to a liquid at the supplier's plant for ease in handling and storage. Conversion of the liquid to a gas for welding use is done by heat exchangers within the cylinder or as a part of the gas delivery system in the welding facility. A Dewar flask and a heat exchanger is shown in Figure 2-29. The following safety rules apply to the Dewar flask system.

    A. Cylinders must always be kept in the vertical position.

    B. Cylinders must always be moved on cylinder carts. These cylinders are extremely heavy and difficult to handle.

    C. Always use the proper equipment when installing or connecting cylinders.

    D. Do not interchange equipment components.

    E. Liquid gases are extremely cold and cause severe frostbite if they come in contact with the eyes or skin. Always wear gloves, safety glasses, and do not touch any frosted surface with bare hands.

3. **Welding Environment Safety.**

    A. Keep the welding area clean.

    B. Keep combustibles out of the welding area.

    C. Maintain good ventilation in the weld area.

    D. Repair or replace worn or frayed ground or power cables.

    E. Make sure the part to be welded is securely grounded.

Figure 2-29. Since a Dewar flask contains very cold liquid gases, a type of heat exchanger is required to change the liquid to a gas. This unit has an auxiliary heat exchanger mounted on the top of the cylinder to increase the amount of liquid conversion to gas.

Figure 2-30. This portable smoke collector is very useful in removing the smoke from the welding operation. The tube and collector can be placed in any position without auxiliary support. (Nederman, Inc.)

    F. Make sure welding helmets have no light leaks.

    G. Use a proper shade number lens to protect the eyes from arc radiation. (See Reference section for the proper lens shade to use.)

    H. Wear safety glasses when grinding or power brushing.

    I. Wear tinted safety glasses when others are tack welding or welding near you.

    J. Use safety screens or shields to protect your work area.

    K. Wear proper clothing. Your entire body should be covered to protect you from arc radiation.

    L. When welding on cadmium-coated steels, copper, or beryllium copper, use special ventilation to remove fumes and vapors from your work area. Figure 2-30 shows a smoke collector near the weld area to remove smoke and vapors. Smoke removal guns can also be used for this purpose.

    M. Do not weld near trichloroethylene vapor degreasers. The arc changes the vapor to a gas. A sweet taste in your mouth indicates that this gas is being formed.

4. **Special Safety Precautions.**

    A. Fires may be started by the welder in a number of ways, such as igniting combustible materials, misuse of fuel gases, electrical short circuits, improper ground connections, etc. Make sure you do not start a fire. If you do, then make sure it is completely out before you leave the area.

B. Never weld on a container that has previously held a fuel until you are sure that it has been purged with an inert gas and tested for fume content.

C. Never enter a vessel or confined space that has been purged with an inert gas until the space is checked with an oxygen analyzer to determine that sufficient oxygen is present to support life.

D. Never use oxygen in place of compressed air. Oxygen supports combustion and will make a fire burn violently.

E. Power brushes are very dangerous, since they eject broken pieces of wire. Always wear safety glasses or safety shields when using this type of equipment.

F. Be alert to the clamping operation when working with mechanical, hydraulic, or air clamps on tools, jigs, and fixtures. Serious injury may result if parts of the body are caught in the clamp.

G. Always know where the fire extinguishers are located.

Review Questions

Write your answers on a separate sheet of paper. Do not write in this book.

1. The electrical flow of current for the GMAW process is usually through the _____ to the electrode.
2. With the electrical flow in the direction described in Question 1, it is always termed _____ or _____.
3. When the welding process is set up so that the electrode is contacting the workpiece and melting off, this is called the _____-_____ mode.
4. Short-arc mode is useful in welding _____ _____ materials because the arc is not on during the entire process.
5. In the _____-_____ mode, the electrode melts off in droplets above the workpiece.
6. Spray-arc mode has a very high metal deposition rate, but is can only be used in the _____ and _____ positions.
7. _____ deposition takes place when the process is set up to produce a large ball on the end of the electrode before melting.
8. The basic welding power for globular deposition is called a constant _____ or constant _____ power supply.
9. Constant current power supplies used in GMAW require a wire feeder with _____ _____ for feeding of the electrode.
10. Inverter-type _____ _____ can be used for all types of GMAW.
11. Some machines have _____ programs that regulate changes in the welding parameters to produce the best possible welding conditions.
12. _____ and _____ are characteristics that are used to establish machine and weld settings for a specific mode of welding.
13. Portable gun-type wire feeders are often called _____ gun systems.
14. The most common gases used with the GMAW process are _____, _____, _____, _____, and _____.
15. Gases used from a high-pressure cylinder are reduced to a lower and workable pressure with a(n) _____.
16. A(n) _____ is used to control the amount of gas delivered to the weld zone through the shielding nozzle.
17. The amount of gas delivered through the gas nozzle is measured in _____ _____ _____ _____ or _____ _____ _____.
18. Too much gas flow through the nozzle causes _____ in the gas flow, which can affect the weld quality.
19. Welding guns are _____-cooled for light duty and _____-cooled for heavy duty welding.

Safety Test Questions

Write your answers on a separate sheet of paper. Do not write in this book.

1. Power supply _____ electrical voltage may range between a low of 115V ac to a high of 440V ac.
2. Contact with this _____ can cause shock and may also cause _____.
3. Never use fuses of a(n) _____ rating than specified.
4. Before welding, make sure the ground clamp is securely attached to the power supply and _____.
5. Install all machines in accordance with the proper _____ code.
6. If you are not a competent electrical technician, never open the _____ _____ cabinet.
7. If you are to work on a power supply, always remove the main power _____ before you work on the unit.
8. The open circuit voltage range of a constant current and a constant voltage power supply is _____-_____ volts.
9. Gas storage cylinders are controlled by the rules of the _____ _____ _____. Older cylinders were controlled by the _____ _____ _____.
10. Shielding gases _____ be detected by the human senses.
11. _____ of the shielding gases are considered toxic.
12. Shielding gas cylinders contain up to _____ psi pressure.

13. Each cylinder must have a(n) _____ listing the contents of the cylinder.
14. All cylinders must be secured with a(n) _____, _____, or a(n) _____ to prevent accidental tipping over of the cylinder.
15. When a cylinder is not in use always keep the protective _____ on the cylinder.
16. Always use a(n) _____ _____ when moving gas cylinders.
17. Always _____ a gas cylinder prior to the installation of the regulator. This removes dirt from the cylinder gas outlet.
18. When opening a cylinder valve, the regulator adjusting screw should always be free of _____ on the regulator diaphragm.
19. Always stand to the side of the pressure gauges when _____ tank pressure valves.
20. When cylinders are not in use, the cylinder _____ should be closed and the diaphragm _____ should be released.
21. _____ return cylinder caps to empty cylinders.
22. Liquefied gas cylinders or bottles are commonly called _____ _____ and they are basically _____ bottles.
23. Dewar flasks are very heavy and must be moved with special _____ carts.
24. Dewar flasks must always be kept in the _____ position to operate properly.
25. Always wear gloves and _____ equipment when working with this system and equipment.
26. Always keep any combustibles _____ of the weld area.
27. Maintain good _____ in the weld area.
28. Make sure your welding _____ has no light leaks.
29. Use the proper welding _____ for the amount of welding current you are using for welding.
30. Make sure your ground is _____ attached to the metal you are welding.
31. Always wear protective _____ or _____ when grinding or power brushing.
32. Wear proper clothing to _____ your body from arc rays and sparks.
33. Wear _____ glasses when others are tack welding or welding near you to prevent arc flashing in your eyes.
34. Use _____ or screens to isolate your area from others who are welding nearby.
35. Be aware of degreasing operations and the possibility of fire. Always know where the _____ _____ are located.
36. Never use _____ in place of compressed air.
37. Never weld on a vessel that has held a fuel until you are sure that all of the _____ are removed.

Welding Safety Checklist

Hazard	Factors to Consider	Precautionary Summary
Electric shock can kill	• Wetness • Welder in or on workpiece • Confined space • Electrode holder and cable insulation	• Insulate welder from workpiece and ground using dry insulation, rubber mat, or dry wood. • Wear dry, hole-free gloves. (Change as necessary to keep dry.) • Do not touch electrically "hot" part or electrode with bare skin or wet clothing • If wet area and welder cannot be insulated from workpiece with dry insulation, use a semiautomatic, constant-voltage welding machine or stick welding machine with voltage-reducing device. • Keep electrode holder and cable insulation in good condition. Do not use if insulation is damaged or missing.
Fumes and gases can be dangerous	• Confined area • Positioning of welder's head • Lack of general ventilation • Electrode types, i.e., manganese, chromium, etc. • Base metal coatings, galvanizing, paint	• Use ventitilation or exhaust to keep air-breathing zone clear and comfortable. • Use helmet and position of head to minimize fume breathing zone. • Read warnings on electrode container and material safety data sheet for electrode. • Provide additional ventilation/exhaust where special ventilation requirements exist. • Use special care when welding in a confined area. • Do not weld unless ventilation is adequate.
Welding sparks can cause fire or explosion	• Containers that have held combustibles • Flammable materials	• Do not weld on containers that have held combustible materials (unless strict AWS F4.1 explosion procedures are followed). Check before welding. • Remove flammable materials from welding area or shield from sparks, heat. • Keep a fire watch in area during and after welding. • Keep a fire extinguisher in the welding area. • Wear fire-retardant clothing and hat. Use earplugs when welding overhead.
Arc rays can burn eyes and skin	• Process: gas-shielded arc most severe	• Select a filter lens that is comfortable for you while welding. • Always use helmet when welding. • Provide nonflammable shielding to protect others. • Wear clothing that protects skin while welding.
Confined space	• Metal enclosure • Wetness • Restricted entry • Heavier-than-air gas • Welder inside or on workpiece	• Carefully evaluate adequacy of ventilation, especially where electrode requires special ventilation or where gas may displace breathing air. • If basic electric shock precautions cannot be followed to insulate welder from work and electrode, use semiautomatic, constant-voltage equipment with cold electrode or stick welding machine with voltage-reducing device. • Provide welder helper and method of welder retrieval from outside enclosure.
General work area hazards	• Cluttered area	• Keep cables, materials, tools neatly organized.
	• Indirect work (welding ground) connection	• Connect work cable as close as possible to area where welding is being performed. Do not allow alternate circuits through scaffold cables, hoist chains, ground leads.
	• Electrical equipment	• Use only double insulated or properly grounded equipment. • Always disconnect power to equipment before servicing.
	• Engine-driven equipment	• Use only in open, well-ventilated areas. • Keep enclosure complete and guards in place. • Refuel with engine off. • If using auxiliary power, OSHA may require GFCI protection or assured grounding program (or isolated windings if less than 5 KW).
	• Gas cylinders	• Never touch cylinder with the electrode. • Never lift machine with cylinder attached. • Keep cylinder upright and chained to support.

It is a very good idea to post a list of safety precautions, like the one shown here, in work areas. Workers should be encouraged to review the safety checklist periodically. (Lincoln Electric Company)

CHAPTER 3

Equipment Set-up and Control

Objectives

After studying this chapter, you will be able to:
- Name four power supply specifications to which the welding machine must conform.
- Describe the effects of adjusting open circuit voltage, arc voltage, slope, and inductance power supply controls.
- List nine typical maintenance guidelines.
- Discuss both major and auxiliary controls of wire feeders.
- Discuss maintenance of the wire feed system.
- Tell how to maintain cables and guns.

Setting up the equipment correctly for the GMAW process is essential. Proper *set-up* not only guarantees intended performance standards, but also ensures that welding work can be done safely without equipment malfunctions.

Power Supplies

Power supplies are specially designed machines that produce welding current for melting the welding electrode at a low voltage. The equipment must be able to control the operation in the areas of:
- Input voltage (primary voltage).
- Open circuit voltage.
- Output ratings and performance.
- Duty cycle.

The National Electrical Manufacturers' Association (NEMA) has established specification *EW-1 Electric Welding Apparatus* for control of these areas.

Power Supply Specifications

Each power supply is designed for specific purposes. Therefore, limitations are established for the proper operation of the machine. The specification areas for machines using utility power include:

- **Primary power type, voltages, and cycles.** This includes alternating current single- or three-phase power at 110, 208, 230, or 460 volts and generally 60 hertz (cycles).
- **Primary power fusing.** The fuse sizes are specified on the machine and in the instruction manual for the individual machine. These limits should never be exceeded.
- **Rated welding amperes.** The rated welding amperes are specified by the machine's manufacturer. These amounts of current should not be exceeded, since the cooling system cannot carry away the excessive heat generated in the machine.
- **Duty cycle.** All welding power supplies are designed to operate for a specific time period at a specific load. The design considerations include:
A. Size of internal wiring.
B. Type of internal components.
C. Insulation of internal components.
D. Amount of cooling required.

The duty cycles range from 20% to 100%. They may be made to a company specification or the NEMA EW-1 specification. This specification establishes that each 10 percentage points represents one minute of operation in a ten-minute period. Figure 3-1 shows the cycle and time period for a NEMA-rated machine.

For example, the duty cycle might rate a 150 ampere welder with a 30% duty cycle to weld at 150 amperes for three minutes. The machine then has to idle for seven minutes. This allows internal components to cool properly before resuming welding.

Machines that are not made to NEMA specifications have various duty cycles and time periods. The duty cycle and time periods should be listed on the machine or in the instruction manual. **Never exceed the equipment manufacturer's duty cycle requirements; doing so will cause failure of the machine components.**

Duty Cycle	Number of Minutes Machine Can Be Operated at Rated Load in a 10 Minute Period
100%	Full Time
60%	6
50%	5
40%	4
30%	3
20%	2

Figure 3-1. Power supply duty cycles limit the number of minutes that a unit can be operated at rated load.

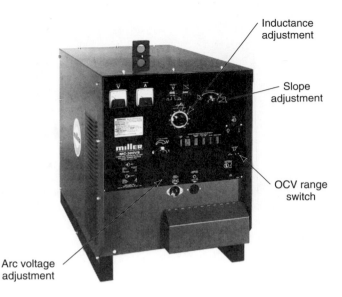

Figure 3-2. This machine is rated at 300 amperes dc with adjustable voltage, inductance, and slope. (Miller Electric Mfg. Co.)

Power Supply Controls

Depending on the type of power supply, the type and number of controls provided for the welder may range from a tap and a simple rheostat to a multitude of controls that vary each function. A small autobody type machine may only have a tap connection for the voltage setting and a wire feed adjustment for wire speed. Another machine may be used for making very high quality welds and would need a number of controls to meet its requirements. The various controls that may be located on a machine, as shown in Figure 3-2, include:

- **Open circuit voltage (OCV) range.** GMAW power supplies produce a range of open circuit voltages to a high near 50 volts. This range of voltages may be controlled by changing taps, switches, or levers that are usually located on the front of the machine. As a general rule, the short-arc mode of welding requires a setting on the low end of the OCV scale and the spray-arc mode requires a setting on the middle to upper end of the scale. When setting up a welding power source, the welder actually sets the ***open circuit voltage.*** After the weld is started, the actual arc or load voltage is established. The OCV may be established on machines that are equipped with voltmeters by following these steps:

1. Turn power supply on.
2. Release idler roller pressure on the wire feeder to prevent wire feeding.
3. Place the voltage range switch in the desired location. (Check machine manual to select range of open circuit voltages desired.)
4. Hold the gun away from ground or workpiece and energize the contactor switch. (When you depress the switch, the contactor allows current to flow to the electrode tip and the wire. An arc will be made if the contact tip or wire touches ground.)

5. Observe the voltmeter and adjust the fine tuning voltage control to the desired open circuit voltage. This is the open circuit voltage; arc voltage will be approximately 2 to 3 volts lower for each 100 amperes.
6. Release the trigger or switch and reset tension on the wire feeder idler roller.

- **Arc voltage adjustment.** Also called ***load voltage,*** the ***arc voltage*** determines the actual arc gap established during the welding period. Adjust the fine voltage control (the same control used to set the open circuit voltage) to the voltage desired during welding.
- **Slope adjustment.** The word ***slope*** in GMAW refers to the slant of the ***volt-ampere curve*** and the operating characteristics of the power supply under load. In many machines, the slant of the volt-ampere curve is automatically set as you change the open circuit voltage. In others, the slope can be changed by the operator for different modes of welding. The curve is known as either *flat* or *steep* mode.

When using a steep slope, as shown in Figure 3-3, there may not be enough current to melt the electrode at the proper time in the short-arc cycle. If the arc will not start properly and the wire stubs out on the workpiece, the slope of the curve has to be decreased (flattened) to operate properly.

A machine set-up with a rather flat slope, as shown in Figure 3-3, has too much current available for the short-arc cycle. As a result, the wire will be blasted off the end, resulting in considerable spatter. In this case, the situation requires changing to a steeper slope.

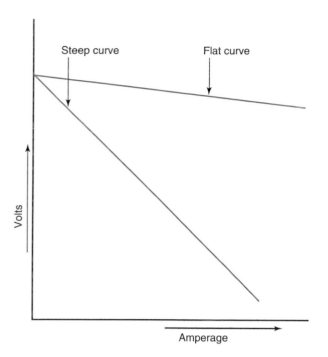

Figure 3-3. The steep curve (slope) setting is normally used for short-arc welding. The flat curve (slope) setting is normally used for spray-arc welding.

When adjusting the slope of the volt-ampere curve on the power supply, always adjust it so that the parting of the molten drop is smooth with a minimum of spatter.

Remember, the main function of the power supply is to provide amperage to maintain the arc gap or voltage selected for welding. For this reason the machine is termed a *constant voltage* (CV) or *constant potential* power supply.

• **Inductance (pinch effect) adjustment.** When using the short-arc mode in GMAW, the separation of the molten drops of metal from the electrode is controlled by the squeezing forces exerted on the electrode due to the current flowing through it. Figure 3-4 shows how the pinch effect operates. This is called **inductance** by some machine manufacturers and the *pinch effect* by others.

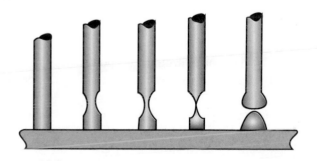

Figure 3-4. As electrical current flows through the electrode, heating takes place until the wire melts off the end. Each cycle is completed many times per second during the welding operation. This is referred to as inductance or the "pinch effect."

By adjusting the level of pinch effect, the machine controls the rate of current rise when the electrode contacts the workpiece and the short circuit starts. If the current flows rapidly through the electrode, the drop of metal is squeezed off quickly and causes spatter. If inductance is added to the circuit and the current is not applied as rapidly, the number of short circuits per second will decrease, and the "arc on" time will increase. This will result in a more fluid weld pool, a smoother weld crown, and less spatter.

In GMAW spray-arc mode, some inductance is beneficial in starting the arc process. This limits explosive starts by slowing down the rate of current rise at the start of the cycle.

As a general rule, both the amount of short circuit current and the amount of inductance needed for the ideal pinch effect are increased as the electrode diameter is increased.

When setting up the machine for and during welding with the short-arc mode, the following guide may be useful:

Maximum inductance (minimum pinch).
• More penetration.
• More fluid weld pool.
• Flatter weld.
• Smoother bead.

Minimum inductance (maximum pinch).
• More convex bead.
• Increased spatter.
• Colder arc.

Preprogrammed Welding Machine Controls

When using a power supply that has built-in programs for various types and thicknesses of materials, select the program desired and insert the required procedure data. This will then automatically set the correct machine values for the operation and monitor the machine output during the welding operation. The machine shown in Figure 3-5 is of this type. The face of the machine has only push buttons for program selection and modification of each value desired. Each readout of the program is displayed on the digital meter. The factory program may also be modified by manual override, if desired. *Note: Each machine will also have areas where programs may be installed and modified by the operator for individual welding procedures. Machines of this type are usually factory programmed for the welding of steel, stainless steel, and silicon bronze or aluminum.*

Pulsers and Auxiliary Pulsers

Some machines have a *pulser system* in the main power supply. Where the power supply does not contain a pulser, an auxiliary machine may be added for pulsing the welding operation. Units like the one shown in

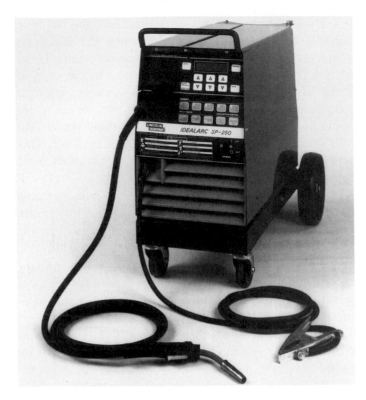

Figure 3-5. This machine has several welding programs already installed by the manufacturer. (Lincoln Electric Co.)

Figure 3-6 are attached to the system and manually set for the desired pulse on and off times. Always refer to the operation manual for a list of set-up options for this process mode and applications. Figure 3-7 shows how the adjustments regulate the pulse on and off times to control the length of the metal transfer period of the weld cycle.

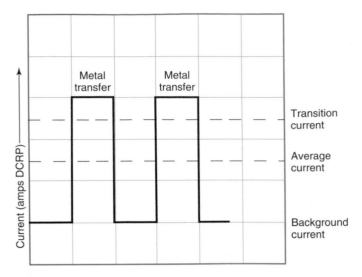

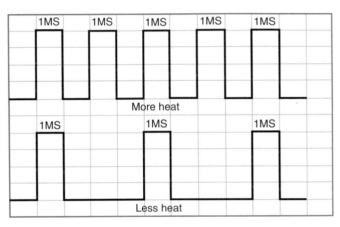

Figure 3-7. Adjusting for a greater number of pulses per second results in more heat and increases metal transfer. Adjusting for fewer pulses per second results in less heat and decreases metal transfer.

Power Supply Installation

A power supply should be installed in an area that is free of dust, dirt, and fumes. There should also be adequate ventilation, so that machine heat can escape. Dirt and heat will cause a power supply to overheat, ruining the internal components. The area selected for installation should be free of objects blocking the flow of air into and out of the machine. The machine should not be exposed to moisture, since electronic components may pick up moisture and fail. Most machine manufacturers require a defined space around the power supply for air circulation; warranties are voided if these areas are not provided.

Utilities supply electrical power at 60 cycles (hertz) and at various input voltages requested by the user. Machines using other than 60 cycle power are specially made at the factory for this requirement. The required fuse size for the incoming power is always shown on the data label. See Figure 3-8. The fuse panels should always be close to the power supply. This makes it possible to

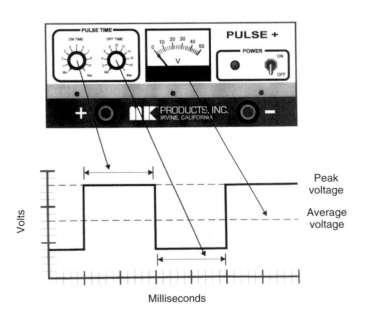

Figure 3-6. Pulser units of this type may be added onto standard CV power supplies. The pulse range is from 30 to 500 pulses per second. (MK Products, Inc.)

disconnect the main power in an emergency. A typical power supply installation is shown in Figure 3-9.

Machines that do not match the voltage supplied by the utility may be used if a step-up or step-down transformer is used. Figure 3-10 shows a step-down transformer installed for this use.

Manufacturers have excellent warranties on their products. Using a power supply voltage or fuses other than the ones stated on the data label or instruction manual may nullify a warranty.

Power Supply Maintenance

Given reasonable care and routine maintenance, a welding power supply will operate for many satisfactory

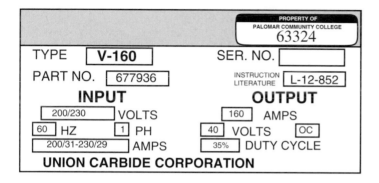

Figure 3-8. The data label lists the requirements for incoming power and the rated output power for the power supply.

Figure 3-10. These transformers change incoming primary power to match the welding equipment's power supply requirements.

hours before repairs are required. Modern methods of insulating transformers, using good basic design and solid-state components, have extended the hours of operation of a modern power supply. The manufacturer's instructions should always be followed for periodic inspection of the power supply. The following points should be observed along with the manufacturer's instructions:

- **Turn the machine off and disconnect power at the fuse box before working on it.**
- Clean or blow out the unit on a periodic basis. Use only dry filtered compressed air, nitrogen gas, or an electrical nonconducting cleaner. **Always wear goggles or a face shield when using compressed air or gas.**
- Check all terminals for loose connections.
- Lubricate fan and motor bearings as required.
- Check mechanical arms and switches for freedom of movement. Mechanical connections may be lightly greased if required.
- Check terminal blocks for cable connections or tap connectors. They should be clean and tight. If corroded, they will restrict the flow of electrical current. They may be cleaned by wire-brushing or by rubbing with an abrasive pad.
- Check motor generator brushes, replacing them when worn beyond the manufacturer's tolerance. Worn brushes wear armatures, which must then be machined for proper operation.
- Lubricate, inspect, and adjust portable gasoline and diesel power supplies, paying close attention to the manufacturer's instructions. This applies to both the engine and the welder. Operation of these units usually occurs under adverse conditions. Improper maintenance will reduce the capacity and operation of the unit.

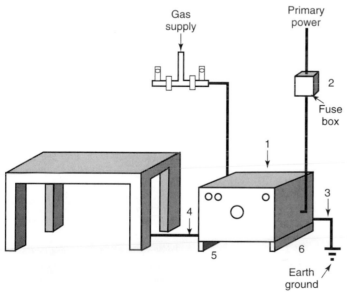

1. Locate power supply away from walls for proper air flow.
2. Locate primary fuse box near power supply.
3. Connect power supply frame to earth ground.
4. Connect work ground with 2/0 min-4/0 max cable.
5. Keep power supply dry.
6. Keep area near power supply clean.

Figure 3-9. A typical power supply installation includes a fuse box mounted near the welder. An earth ground from the power supply protects the welder from primary line high voltage.

- Use extreme caution when fueling gasoline or diesel engines.
- Replace components only with authorized replacement parts.

Wire Feeders

Most of the feeders use 110V ac power which is provided to the machine by a connection in the power supply. If this connection is used, the wire feeder is turned on or off when the power supply is operating or not operating.

Wire feeders that are installed on push-pull systems or in hand-held guns use 24V dc motors for precision drive of the wire and safety of the welder.

Types of Wire Feeders

The basic types of wire feeders used for GMAW include:

- **Push-type.** This method is used to push the wire from the spool to the welding gun. A standard feeder is shown in Figure 3-11.
- **Pull-type.** This method is used to pull the wire from the supply spool to the drive rollers in the welding gun. Refer to Figure 2-20 for an example of a pull-type gun.
- **Push-pull type.** This method is used to push the wire from the wire feeder to a set of drive rollers mounted in the welding gun. This type of system

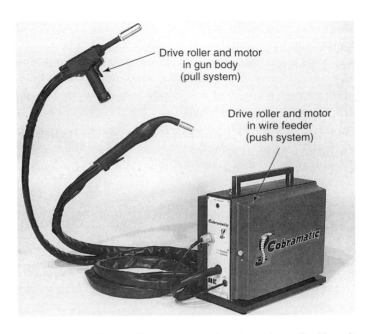

Figure 3-12. Push-pull systems work extremely well with soft wires, even when the wire must be moved a considerable distance from the wire feeder. (MK Products, Inc.)

is used for welding with soft or small-diameter wires, since these wires may buckle in the cable liner if pushed long distances. A unit of this type is shown in Figure 3-12.

- **Spool gun type.** This type of unit has a spool of wire located on the welding gun. Since the size of the wire supply spool is small, the amount of welding is limited. A spool gun is shown in Figure 3-13.

Wire Feeder Controls

The number and types of controls included in the wire feeder will vary depending on the use of the feeder and the amount of wire feed desired. See Figure 3-14.

Figure 3-11. Wire feeders of this type are called push-type feeders because they push the wire to the welding gun. (Miller Electric Mfg. Co.)

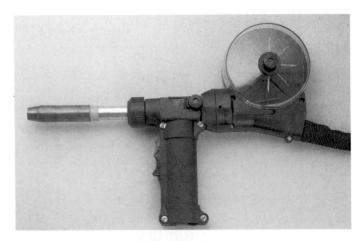

Figure 3-13. The trigger activates and stops the welding sequence, while a rheostat mounted on the gun controls the wire speed. (Tweco-Arcair, a Thermadyne Company)

Figure 3-14. Manufacturers place various controls on wire feeders, depending on the use of the machine. (Lincoln Electric Co.)

The major controls include the power switch (off/on control), the *wire feed potentiometer* (speed control), and the *spool brake control* (stops wire spool at end of welding).

Auxiliary wire feeder controls may include:

- **Mode switch.** This control is used to select various modes, such as spot, intermittent, and seam welding.
- **Trigger lock-in control.** This control allows the welder to weld without the need to depress the operating switch during the entire operation.
- **Burnback (anti-stick) control.** This control prevents the wire from sticking in the weld pool. It sets the time that the arc power is on after the stop switch is released to burn back the wire from the molten weld pool. This control must be used when spot welding.
- **Spot weld time.** This control sets the length of time for the spot weld operation.
- **Stitch weld time.** This control sets the time needed to make a long seam or a specific weld length.
- **Prepurge timer control.** This control establishes the time that the gas will flow before welding will start.
- **Purge control.** This control allows the welder to open the gas solenoid. The amount of gas flow can be set on the flowmeter, or the welder can purge the gas lines prior to welding.
- **Wire inching (jog) control.** Moves the wire through the wire feeder to the contact tip. It also may be used by the welder when determining wire speed. (Jog speed is the same speed as the set wire speed.)

- **Wire reel brake control.** The reel brake control should be adjusted so that the spool will stop rotating at the same time the wire stops feeding at the end of the welding operation. If the wire is allowed to overspool and touch the feeder housing, a short circuit may occur (if the electrode circuit is energized), or the wire may *birdnest* (tangle). The brake adjustment is a tension device located in the hub of the wire spool. When changing reels of wire, Figure 3-15, always replace the reel locking pin, and check the brake adjustment.

Wire Straightener

Wire straighteners are placed in the supply system to reduce the amount of *wire cast* (curvature) before the wire is drawn into the drive roller system. Excessive cast is often found on hard wires, causing heavy wear of the contact tips during welding. This can result in *arc outages* (loss of arc). Increasing or decreasing the amount of cast is done with an adjusting wheel on the straightener. A three-roll wire straightener is shown in Figure 3-16.

Wire Feeder Drive Roll System

The system for driving the electrode wire from the spool to the gun is either a two-roll or a four-roll system. A standard two-roll system is shown in Figure 3-17, and a four-roll drive system is shown in Figure 3-18. The rollers shown in Figure 3-19 are grooved for driving specific sizes and types of wire. If the groove does not match the wire type and size, the wire will seize in the groove and not feed properly.

Figure 3-15. Adjusting the brake increases or decreases tension on the wire reel hub to prevent overspooling of wire.

Figure 3-16. Adjusting the pressure roller will increase or decrease the amount of cast in the wire. (Miller Electric Mfg. Co.)

Figure 3-17. Two-roll wire feed system. Note the feed roll has two grooves to adapt to different wire sizes. (CK Worldwide, Inc.)

The idler roller part of the system may be smooth or have grooves for the specific wire type and diameter. The idler roller must be adjusted to feed the wire into the roller grooves without slippage or flattening the wire.

Inlet guides are used to feed the wire into the drive rollers. Outlet guides are also used to feed the wire from the drive rollers into the gun cable or cable adapter. Guides are made of a variety of materials. Hard wires (steel, stainless steel, Inconel, etc.) will wear the guides quite rapidly and require frequent replacement. They are adjusted as close to the roller as possible. This prevents the wire from "birdnesting" in the area between the rollers and guides. See Figure 3-20.

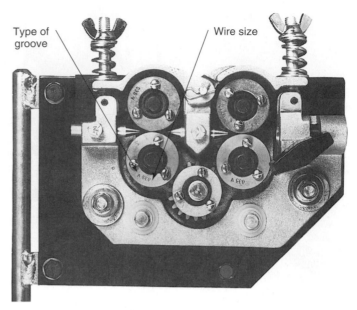

Figure 3-18. Four-roll wire feed system. These systems are used where precision drive of the filler wire is required. Note that rollers have the groove type and wire size stamped on them. (Miller Electric Mfg. Co.)

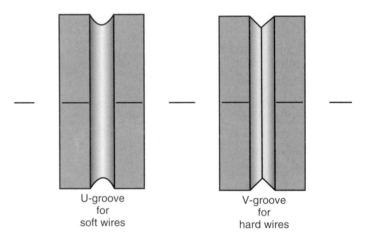

U-groove for soft wires

V-groove for hard wires

Figure 3-19. Wire drive rollers have different groove designs for hard or soft wires. Using the incorrect groove will cause wire feeding problems.

Wire Wipers

This accessory is attached to the inlet side of the drive roll system, as shown in Figure 3-21. It usually consists of a piece of felt material that is closed where the wire passes through it. A small amount of commercial fluid may be placed on the felt to lubricate the wire as it passes through. This will extend the life of the guides and the cable liners in the gun cable. Use the wiper only when welding on the specific materials defined by the manufacturer of the fluid. Do not over-lubricate.

Wire Feeder Maintenance

Wire feeders require very little maintenance to keep them operating properly. As with any electrical device,

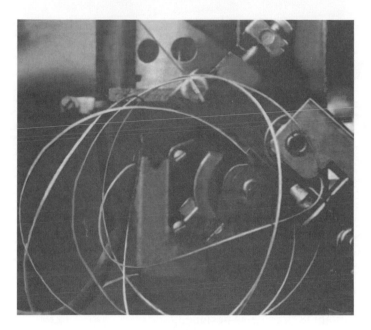

Figure 3-20. "Birdnesting" of the wire is often caused by improper operation of the wire feed system. Check the guides, rollers, and tension to find the problem and correct it.

they must be kept dry to protect the electrical components within the feeder. Electrical malfunctions should be checked only by a skilled electrician. The printed circuit boards (if the feeder is so equipped) should be replaced only with factory authorized parts obtained from the welding supplier.

The major problems will be found in the drive roll system and the guides to and from the rollers. Worn equipment will cause many problems in the system. Inspect the guides and rollers often and replace when worn. If you have feeding problems, the wire feed mechanism is most likely where the problem exists. To determine if the feeder is at fault, disconnect the gun cable and run the wire through the feeder only. The wire should run through smoothly. If the feeder is operating

satisfactorily, connect the gun and cable to the feeder. Operate the feeder with the gun and cable attached. If the fault remains, the problem is in the gun and cable assembly. Be sure you know where a problem is before you try to repair it.

Cables and Guns

Cables

Cables are used to carry electrical current, shielding gas, and welding wire to the gun. In some cases, the cable includes tubing to carry cooling water to and from the gun. Some cables will also include a circuit wire from the switch to the machine contactor, which is used to start and stop the weld operation. The cables are usually made in specified lengths. Some may be connected to other sections to lengthen the cable.

Where the cable attaches to the wire feeder, special adapters are necessary. As a general rule, each type of gun requires an adapter that is made by the gun manufacturer. A cable adapter is shown in Figure 3-22.

To protect the cables and guns from wear as the wire passes through them, *liners* are installed. In some cases, special liners are installed in the gun. Manufacturers will specify which liner must be used with each type and size of wire. Some wires may require use of a special lining material, such as Teflon® or nylon. The cable liner shown in Figure 3-23 is always installed into the gun cable from the feeder end and secured at the welding tip end. See Figure 3-24. Since each manufacturer makes guns to a specific design, the liners are not interchangeable. When changing spools and types of wire, always blow out the liner with clean, dry compressed air to remove any residue from the previous wire.

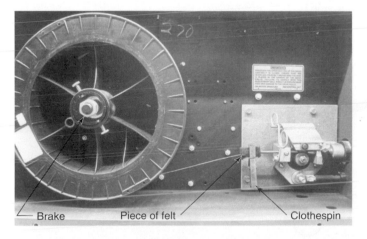

Brake   Piece of felt   Clothespin

Figure 3-21. A wire wiper can be assembled from a clothespin and a piece of felt. Commercial fluids may be used in the wiper material. (Ferro-Slick/Lube-Matic)

Gas solenoid

Gun adapter

Figure 3-22. Adapters can be used to connect different types of guns to a wire feeder.

Guns

Guns are designed to carry current with a specific duty cycle, just as welding machines are. They are usually rated by the type of gas that is being used when they are gas-cooled. Guns that are used for large-diameter wires are usually water-cooled for a higher duty cycle. A water-cooled system is shown in Figure 3-25.

If water from a municipal system is used for cooling, as shown in Figure 3-26, the system requires a filter and a regulator to match water pressure to the manufacturer's requirements.

These additions are needed because municipal water often contains small particles that could clog gun cooling passages and ruin the welding torch. For this reason, a filter must be cleaned or replaced often to prevent reduced water flow. Manufacturers usually specify cooling water pressure of no more than 50 psi.

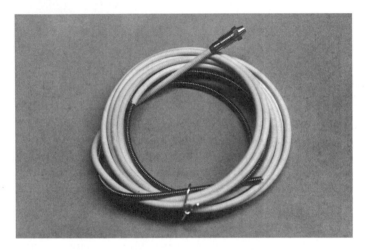

Figure 3-23. A new gun liner, ready for installation.

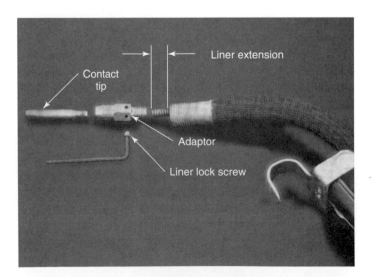

Figure 3-24. Gun liners must fit into the adapter with a specific extension. Incorrect assembly of this part will cause wire feeding problems. Always follow the manufacturer's instructions when installing liners. (Tweco Products, Inc.)

Since municipal water pressure varies, a regulator must be installed to prevent too much pressure in the cooling hoses or the gun. Failure to filter the water and regulate its pressure can result in clogged guns, burst hoses, and possibly damage to the welding gun.

Spool Guns

Guns of this type are designed for welding aluminum or other soft alloy wires. All of the types of guns are rated for a specific duty cycle. They contain a low voltage motor and drive rollers for moving the electrode from the spool to the contact tip, Figure 3-27. Electrode speed is controlled by a rheostat and a trigger to start and end the sequence. The drive and idler rollers are of various designs. For this reason, when changing types of wire, diameters, or replacing parts, the rollers must be of the same model and manufacturer for the system to operate as designed.

Gun Parts

As shown in Figure 3-28, the end of the gun has several components. They are used to receive the liner assembly, insulate the gas nozzle, and transfer electrical current to the contact tip and the electrode. These

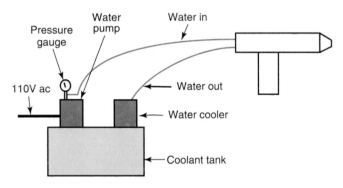

Figure 3-25. This type of a unit is called a closed system. It eliminates many of the problems associated with water-cooling the gun.

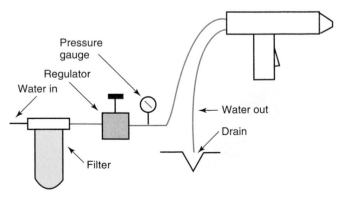

Figure 3-26. Use of water from municipal systems requires more equipment to do an efficient job of cooling without wasting water.

Figure 3-27. This unit is very safe for the operator to use, since a very low voltage motor is used to drive the wire. (Miller Electric Mfg. Co.)

components should fit together firmly for the proper operation of the gun. Insulators that are cracked or burned should be replaced.

Contact tips are designed for operation with a certain type and diameter of filler material and a specific welding mode. Each mode of welding will define the relationship of the contact tip end with the end of the gas nozzle. As a general rule, the short contact tips are used with the spray-arc mode, as shown in Figure 3-29; the longer contact tips, as shown in Figure 3-30, are used with the short-arc mode. Various types of contact tips are shown in Figure 3-31.

When the exit hole size becomes enlarged to the point where electrical contact with the wire cannot be maintained, the tip must be replaced.

Continued use of an oversize hole in the contact tip will cause arc outages, incorrect heating of the wire, cold laps in the weld pool, and poor weld quality.

Always use the manufacturer's recommended contact tip for the job involved and inspect regularly to make sure the tip is seated firmly into the adapter seat. Since the tips are expendable, it is suggested that several tips for each application be kept in stock.

Gas nozzles are made by each manufacturer to fit an individual gun, and they are not generally

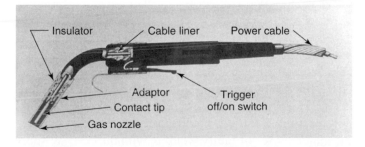

Figure 3-28. The components of a typical gun are shown in the cutaway view.

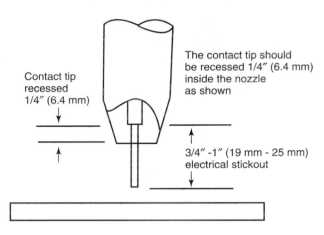

Figure 3-29. Spray-arc contact tips fit within the nozzle area.

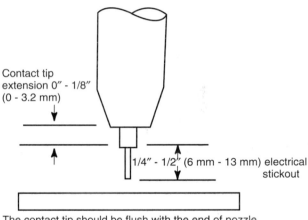

The contact tip should be flush with the end of nozzle or extend a maximum of 1/8″ (3.2 mm) as shown.

Figure 3-30. Short-arc contact tips fit flush with the nozzle end or extend beyond the nozzle tip end.

interchangeable. However, manufacturers generally use the same system for dimensioning the exit size of the nozzle end. The exit diameter is usually dimensioned in sixteenths of an inch. This is shown in Figure 3-32. In some cases, where high currents are carried, the gas nozzle will get extremely hot causing the copper to peel or flake. To correct this condition, some nozzles have aluminum fins on the outside to radiate heat away from them. A finned nozzle is shown in Figure 3-33.

During the welding operation, spatter from the molten metal will gather on the inside of the nozzle, as shown in Figure 3-34. Such contaminated nozzles decrease and deflect the gas flow over the molten metal, causing oxidation of the weld metal. To prevent this condition, use an anti-spatter compound or spray to coat the inside of the nozzle. An anti-spatter spray is shown in Figure 3-35. A brush or a tool like the one in Figure 3-36 also may be used to remove spatter buildup between spraying applications.

Figure 3-31. Manufacturers design contact tips for individual guns.

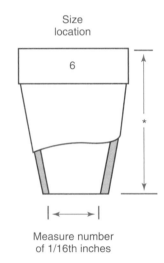

Size		No.
3/16	=	No. 3
4/16	=	No. 4
5/16	=	No. 5
6/16	=	No. 6
7/16	=	No. 7
8/16	=	No. 8
9/16	=	No. 9
10/16	=	No.10
11/16	=	No.11
12/16	=	No.12

Lengths*
Short
Regular
Long
Extra long
Special

Size location

6

Measure number of 1/16th inches

Figure 3-32. Standard nozzle diameters and lengths.

Figure 3-33. Air-cooled nozzles are used to carry heat away from the nozzle end during high-amperage welding. (MK Products, Inc.)

Figure 3-34. Contaminated nozzles like these will prevent sufficient gas flow over the weld. Faulty welds will result.

Gun Maintenance

The welding gun is designed to operate properly if it is maintained with reasonable care. Treating the gun in a rough manner will damage the outer case and may cause internal electrical shorts which could ruin the gun. Do not force threaded assemblies together or use parts designed for other guns. When changing adapters or contact tips, use a wire brush to clean threaded areas to assure good electrical contact. Check liner set screws often and use *dry* compressed air or an inert gas to blow out the liner assembly. Plastic or steel liners used in the front of the gun assembly should be inspected occasionally for wear and possible replacement.

Guns that use a pull system or a wire feeder with segmented teeth should be checked often for proper wire tension. The adjustment screw should be tight enough to pull the wire without permitting the teeth of the wheel to excessively mark the wire. Excessive marking or indentation will cause flaking of the filler wire and rapidly wear the front liner of the gun.

Ground Clamps (Work Leads)

The importance of proper *ground leads* and proper clamps cannot be overemphasized in GMAW. The cable must be of a sufficient size to adequately carry the welding current without overheating. The cable sizes shown in Figure 3-37 are the minimum sizes; larger cables can always be used. The ground clamp must fit tightly to the workpiece or an arc will be made when the current flows through the connection. Arc outages may occur where good contact is not made; the weld joint at these points may not be satisfactory. Cables that are cracked or broken must be replaced. Connections between the clamp and the cable should be tight, since loose connections will not allow proper current flow. To check the ground system, hold your hand near the cable and connections. If the cable or connections are warm or hot, they require servicing or replacement. This is one of the major causes of faulty welding with the GMAW short arc mode of welding. When circular parts mounted on positioners are being welded or the parts are rotated, a

Figure 3-35. Anti-spatter spray is used to prevent spatter buildup on the end of the nozzle. This should be applied often to the end of the nozzle and the contact tip. (G.S. Parsons Co.)

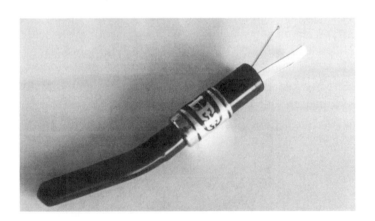

Figure 3-36. Special tools like this one are used to remove spatter from the nozzle tip.

	Welding Current (amps.)	Cable Size (No.)
Recommended Minimum Cable Sizes	100	4
	150	2
	200	2
	250-300	1/0
	300-450	2/0
	500	3/0
	600	4/0

Figure 3-37. Cables should be inspected regularly for cuts or for damage from excessive heating. Always replace with proper size, as shown on this chart.

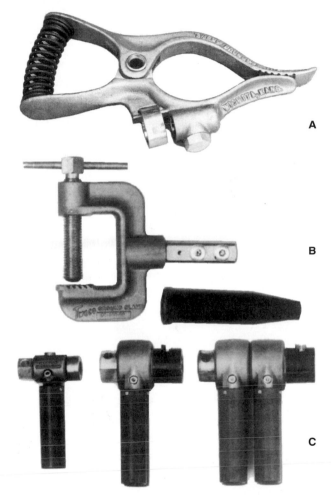

Figure 3-38. Many different types of ground clamps are available for use with GMAW. They must work properly and have clean points for contact with the part being welded. A—Spring-type clamp. B—Screw-type clamp. C—Rotary ground. (Tweco Products, Inc.)

rotary ground may be used. This eliminates tangling of the ground cable and assures good ground contact throughout the welding operation. Various types of grounds for stationary and rotary weldments are shown in Figure 3-38.

REVIEW QUESTIONS

Write your answers on a separate sheet of paper. Do not write in this book.
1. The power supply produces welding current at _____ voltages.
2. Welding power supplies are made to operate on _____ or _____ phase primary current.
3. All welding power supplies have a _____ output current which is established by the design of the machine to _____ specifications.
4. The time at rated load at which the machine will operate without harm to the machine is called the machine _____ _____.
5. This time period is based on a _____ -minute period.
6. The letters OCV with regard to welding power supplies define the _____ _____ _____ established on the machine.

7. The actual arc voltage established on the machine after welding begins may also be referred to as _____ voltage.

8. When welding with the short-arc mode, the machine OCV is generally set in the _____ end of the range, while the spray-arc mode requires setting on the _____ voltage range.

9. The _____ of the volt-ampere curve is actually the operating characteristics of the welding machine output.

10. When setting up a machine for short-arc welding and slope adjustment is available, the slope is set in the _____ setting.

11. When setting up a machine for spray-arc welding and slope adjustment is available, the slope is set in the _____ mode.

12. Another name for inductance is _____ _____. It relates to the rate of rise of welding current when the electrode is in contact with the workpiece in the short-arc mode.

13. Adjustment of the inductance control aids in obtaining _____, reducing _____, and obtaining a _____ weld crown.

14. The word _____ is used to express the numbers of cycles of input electrical current.

15. Wire feeders generally use 110 volts input current while gun motors will have a _____ voltage for the safety of the welder.

16. The three types of wire feeders include the _____-type, the _____-type, and the _____-_____ type.

17. Wire feeders have either a _____ - or _____ -roll drive system. The _____ -roll system is used where precision drive of the filler wire is required.

18. A welding machine manufacturer's guarantee on equipment is called the machine _____.

19. Proper _____ of welding equipment will extend the life of the equipment for a long period of time and reduce the number of major repairs.

20. The gun cable is connected to the wire feeder by the use of an _____. This equipment is usually made by the gun manufacturer and only fits the manu-facturer's equipment.

21. Guns are rated by _____ _____, in the same way as power supplies are rated.

22. The electrical current produced by the power supply is transferred to the electrode through the _____ _____.

23. A _____ is used in the gun cable to protect it from wear and is made to fit a specific _____ of wire. It is installed in the cable with specific instructions for assembly into the gun adapter.

24. Welding guns which use the push-pull system use wheels with _____ _____ that pull the wire from the wire feeder slave unit.

25. Arc outages in the short-arc mode are generally caused by a faulty _____ connection.

CHAPTER 4

Shielding Gases and Regulation Equipment

Objectives

After studying this chapter, you will be able to:

- Explain what functions shielding gases serve in GMAW.
- Compare characteristics of welds made with different gases or gas combinations.
- Tell which gases or gas combinations are used on ferrous and nonferrous metals.
- Find leaks in a gas distribution system.
- Discuss how shielding gases are supplied and distributed in a GMAW system.

Shielding Gases

A *shielding gas* may be a pure gas or a mixture of several gases. In GMAW, gases are used to:

- Shield the electrode and the molten metal from the atmosphere.
- Transfer heat from the electrode to the metal.
- Stabilize the arc pattern.
- Aid in controlling bead contour and penetration.
- Assist in metal transfer from the electrode.
- Assist in the cleaning action of the joint and provide a wetting action.

The common shielding gases include argon, helium, carbon dioxide, and oxygen. Argon and helium are *inert gases* (those that are chemically inactive and will not combine with any product of the weld area). They are pure, colorless, and tasteless. They may be used as a single gas or a part of a mixed gas combination. Carbon dioxide is not inert, but may be used alone or as part of a mixed gas combination. Oxygen is always combined with other gases.

Argon

Argon is the shielding gas most commonly used for GMAW. It is separated from the atmosphere during the

production of oxygen and is, therefore, readily available at low cost.

Argon produces narrow bead widths, because the arc is more concentrated than that of any other gas. The weld penetration is deep in the center of the weld. Since argon gas is heavier than air and tends to form a blanket around the electrode and molten metal, spatter and contamination are reduced.

Helium

Helium is found in natural gas wells and is higher in cost than argon. It is not often used as a single gas in GMAW, because of poor metal transfer from the electrode to the weld pool. Although helium has a higher thermal conductivity than argon, penetration is wider and not as deep. Since helium is lighter than air, the gas tends to rise from the weld area quite rapidly. This makes it necessary to use a higher flow rate to maintain proper shielding.

Carbon Dioxide

Carbon dioxide is a compound gas made up of carbon monoxide and oxygen. These individual gases are mixed, then compressed and stored in liquid form. GMAW uses only carbon dioxide that has had the moisture removed during processing. It is then termed *welding grade carbon dioxide.*

Oxygen

Oxygen is acquired from the atmosphere. It is not used as a single gas for welding, but as part of a gas mixture to attain specific arc patterns. Since the amount of oxygen used is a very small percentage of the total gas mixture, the user should purchase the mixture from a supplier.

Gases and Mixes for Welding Ferrous Metals

- 100% carbon dioxide may be used for short-arc welding of steel.

- 99% argon with 1% oxygen may be used only for welding stainless steel in the spray-arc mode in the flat and horizontal positions. Adding oxygen to argon stabilizes the arc while improving the appearance of the weld bead. Groove welding is limited to the flat position. Fillet welds may be done in either flat or horizontal positions.
- 98% argon with 2% oxygen may be used in the spray-arc mode for welding carbon steels and stainless steels. Groove welding is limited to the flat position. Fillet welds may be done in either flat or horizontal positions.
- 95% argon with 5% oxygen may be used for spray-arc welding carbon and stainless steels. Groove welding is limited to the flat position. Fillet welds may be done in the flat or horizontal positions.
- 75% argon with 25% carbon dioxide is the most common mixture for short-arc welding carbon steels in all positions. Welds made with this gas mixture have minimum spatter and good penetration features.
- 50% argon with 50% carbon dioxide offers many of the qualities of the 75% argon/25% carbon dioxide mixture, but at a lower cost. This is because a smaller amount of the more expensive argon gas is used. When spatter and a deeper penetration can be tolerated, this gas mixture may be used.

Gases and Mixes for Welding Nonferrous Metals

- Argon is used for spray-arc welding in all positions on aluminum, nickel-based alloys, and reactive metals. It may also be used for welding some thin-gauge materials in the short-arc mode.
- Helium gas alone has only limited use in GMAW. Its basic application is in machine welding of heavy aluminum at high welding currents.
- 75% argon with 25% helium is normally used on heavier materials with the spray-arc mode. Penetration is deeper than with pure argon and the weld bead appearance is good.
- 75% helium with 25% argon is used with spray-arc welding of heavier materials. This percentage of helium increases the heat input, which reduces internal porosity and provides good wetting of the weld into the parent metal.

Special Gas Mixtures

- 90% helium with 7 1/2% argon and 2 1/2% carbon dioxide is a mixture developed for short-arc welding of stainless steels. Welding may be done in all positions. The argon addition provides good arc stability and penetration. The high helium content provides heat input to overcome the sluggish nature of the weld pool in stainless steel.
- 60% helium with 35% argon and 5% carbon dioxide is a mixture for short-arc welding of high-strength steel in all positions.

Specialty Gas Mixtures

Some specialty gas mixtures (such as "STARGON" and "MIG-MIX") have been developed by gas suppliers for use in standard and specialized applications. These mixes often allow a broader range of welding parameters than the standard mixes and thus reduce welding costs.

Selecting the Proper Gas Mixture

When selecting the proper gas or gas mixture, you must consider a number of factors:

- Mode of metal transfer.
- Base metal type.
- Base metal thickness.
- Joint design.
- Weld position.
- Filler material composition.
- Filler material size.
- Chemical composition of the desired weld metal.
- Weld metal quality.

The various gases used in GMAW, and their applications, are listed in Figure 4-1A. The gases for GMAW with short circuiting transfer metal deposition are listed in Figure 4-1B. Figure 4-1C lists the selection of gases for GMAW with spray transfer.

Purge Gas Applications

When purging on the root side of the weld is desired or necessary to protect it from atmospheric contamination, always use argon, nitrogen, or helium. Where the application requires very good protection, always use argon or helium. In many cases, however, nitrogen may be used as an inexpensive initial *purge gas* before using the more expensive gases. The nitrogen will displace a large percentage of atmospheric oxygen at a reduced cost. The gas may also be used to protect the back side of fillet or groove welds that do not require full penetration of the root side of the weld. Figure 4-2 shows applications where nitrogen may be used as a purge gas.

Gas Purity

Inert shielding gases used for welding are refined to high purity specifications. Cylinder argon has a minimum purity of 99.996% and contains a maximum of about 15 parts per million moisture (a dew point temperature of −73°F maximum). Driox® brand argon has a minimum

Shielding Gas	Chemical Behavior	Typical Application
Argon	Inert	Virtually all metals except steels.
Helium	Inert	Aluminum, magnesium, and copper alloys for greater heat input and to minimize porosity.
Ar + 20-80% He	Inert	Aluminum, magnesium, and copper alloys for greater heat input and to minimize porosity (better arc action than 100% helium).
Nitrogen		Greater heat input on copper (Europe).
Ar + 25-30% N_2		Greater heat input on copper (Europe); better arc action than 100 percent nitrogen.
Ar + 1-2% O_2	Slightly oxidizing	Stainless and alloy steels; some deoxidized copper alloys.
Ar + 3-5% O_2	Oxidizing	Carbon and some low alloy steels.
CO_2	Oxidizing	Carbon and some low alloy steels.
Ar + 20-50% CO_2	Oxidizing	Various steels, chiefly short circuiting mode.
Ar + 10% CO_2 + 5% O_2	Oxidizing	Various steels (Europe).
CO_2 + 20% O_2	Oxidizing	Various steels (Japan).
90% He + 7.5% Ar + 2.5% CO_2	Slightly oxidizing	Stainless steels for good corrosion resistance, short circuiting mode.
60 to 70% He + 25 to 35% Ar + 4 to 5% CO_2	Oxidizing	Low alloy steels for toughness, short circuiting mode.

A

Metal	Shielding Gas	Advantages
Carbon steel	75% argon + 25% CO_2	Less than 1/8 in. (3.2 mm) thick: high welding speeds without burn-thru; minimum distortion and spatter.
	75% argon + 25% CO_2	More than 1/8 in. (3.2 mm) thick: minimum spatter; clean weld appearance; good puddle control in vertical and overhead positions.
	CO_2	Deeper penetration; faster welding speeds.
Stainless steel	90% helium + 7.5% argon + 2.5% CO_2	No effect on corrosion resistance; small heat-affected zone; no undercutting; minimum distortion.
Low alloy steel	60-70% helium + 25-35% argon + 4-5% CO_2	Minimum reactivity; excellent toughness; excellent arc stability, wetting characteristics, and bead contour; little spatter.
	75% argon + 25% CO_2	Fair toughness; excellent arc stability, wetting characteristics, and bead contour; little spatter.
Aluminum, copper, magnesium, nickel, and their alloys	Argon & argon + helium	Argon satisfactory on sheet metal; argon-helium preferred on thicker sheet material (over 1/8 in. [3.2 mm]).

B

Figure 4-1. A—GMAW gases and applications. B—GMAW short arc gases.   *(Continued)*

purity of 99.998% and a moisture content of less than six parts per million. Helium is produced to a minimum purity of 99.995%. Generally, helium contains less than 15 parts of moisture. At these levels of refinement, impurities usually cannot be detected during welding.

Most steels and copper alloys show high tolerances for various amounts of contaminants. Aluminum and magnesium are sensitive to gas purity and will exhibit severe porosity if contaminated gas is used. Still other materials, such as reactive metals, have extremely low tolerances for contaminants in the inert gases. High purity standards are maintained by the gas suppliers to ensure that shielding gases will be more than adequate for the most severe application.

Gas Supply

Shielding gases are supplied in cylinders of various sizes for shop use, Figure 4-3. *Dewar flasks* (insulated, pressurized containers for liquefied gas) and high-pressure gas cylinders mounted on trailers are used where large volumes of gases are required. See Figure 4-4.

In most cases, gas is sold through a distributor by the total quantity in cubic feet of each type of gas. Since gas is distributed in cylinders, the distributor will also charge a *demurrage* (rental) fee on each cylinder used. When large quantities of gases are needed, the liquefied form is the most economical, since fewer cylinders are used.

Metal	Shielding Gas	Advantages
Aluminum	Argon	0 to 1 in. (0 to 25 mm) thick: best metal transfer and arc stability; least spatter.
	35% argon + 65% helium	1 to 3 in. (25 to 76 mm) thick: higher heat input than straight argon; improved fusion characteristics with 5XXX series Al-Mg alloys.
	25% argon + 75% helium	Over 3 in. (76 mm) thick: highest heat imput: minimizes porosity.
Magnesium	Argon	Excellent cleaning action.
Carbon steel	Argon + 1-5% oxygen	Improves arc stability; produces a more fluid and controllable weld puddle; good coalescence and bead contour; minimizes undercutting; permits higher speeds than pure argon.
	Argon + 3-10% CO_2	Good bead shape; minimizes spatter; reduces chance of cold lapping; cannot be used out of position.
Low-alloy steel	Argon + 2% oxygen	Minimizes undercutting; provides good toughness.
Stainless steel	Argon + 1% oxygen	Improves arc stability; produces a more fluid and controllable weld puddle, good coalescence and bead contour; minimizes undercutting on heavier stainless steels.
	Argon + 2% oxygen	Provides better arc stability; coalescence, and welding speed than 1 percent oxygen mixture for thinner stainless steel materials.
Copper, nickel, and their alloys	Argon	Provides good wetting; decreases fluidity of weld metal for thickness up to 1/8 in. (3.2 mm).
	Argon + helium	Higher heat inputs of 50 & 75 percent helium mixtures offset high heat dissipation of heavier gases.
Titanium	Argon	Good arc stability; minimum weld contamination; inert gas backing is required to prevent air contamination on back of weld area.

C

Figure 4-1. C—GMAW spray arc gases. *(Continued)*

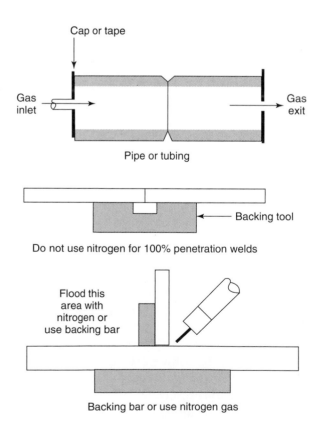

Figure 4-2. Nitrogen applications as a purge gas. Use only on carbon steels.

Gas Storage

Storage of shielding gas cylinders and containers should be carefully controlled to prevent incorrect use. **The gas cylinder or container should always be stored outside or in a well-ventilated area. Cylinders should be stored upright (in a vertical position), with the safety cap in place, and secured to a rigid support to keep them from falling over. The type or mixture of the cylinder's gas contents should be properly identified. Follow all safety precautions to avoid injury. Remember, inert gases do not contain oxygen, and therefore, will not support life. You cannot see, smell, or taste inert gases.**

Gas Distribution

Gases may be distributed to the welding area in several ways. Figure 4-5 shows a bank of cylinders that have been connected together. They may be located in a convenient place with the gas being piped to the welding area. With this arrangement, one bank may be used until empty, and then another can be placed into use. Empty cylinders, in turn, can be replaced without affecting the bank in use.

Manifolds are often used to distribute gases to the welding area from the supply area. Through the use of

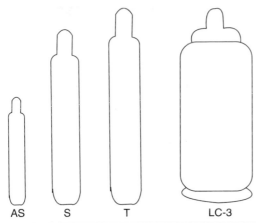

Cylinder Style	Contents Cubic Ft.	Full Pressure of Cylinder at 70° F	Height	O.D.
AS	78	2200	35	7 1/8
S	150	2200	51	7 3/8
T	330	2640	60	9 1/4
LC-3	2900	55	58	20

Figure 4-3. Types and capacities of cylinders and Dewar flasks supplied to industry.

manifolds, the number of individual cylinders required at the welding station can be reduced. The manifold in Figure 4-6 can be used to supply up to six stations. Figure 4-7 shows a high-pressure commercial manifold used to connect several banks together.

The distribution system must be leakfree to maintain the purity of the gas. Therefore, all of the cylinder fittings

Figure 4-5 Individual cylinders are connected in banks to a supply manifold by high-pressure tubing. Empty cylinders can be replaced without interrupting service.

Figure 4-4. Two Dewar flasks containing argon are connected to a pressure regulator and a switching valve.

must be cleaned before installation and properly seated into the regulators. High-pressure connectors, tubing, and pipe connectors must be protected to prevent entry of foreign materials, water, or oil when the system is not in use. Threaded plastic adapters or tape may be placed over any unused openings for this purpose.

Testing for Gas Leaks

Before a system can be placed into use, it must be tested for leaks at a pressure higher than that used as the ***normal operating pressure***. One method is to use a special leak test solution, Figure 4-8. While the pipe is under pressure, the solution is applied. Since the solution is sensitive to gas flow, it will form bubbles wherever there is a leak.

Another method is to apply pressure to the system and note the test pressure. The inlet pressure valve is then closed and the pressure gauge observed closely. Any leak that is present will be indicated by a drop in pressure.

Prior to placing the manifold in use, the lines must be ***purged*** to remove any flux vapors and moisture. This can be done by using nitrogen gas as the initial purging

Figure 4-6. Dewar flasks supply gas at approximately 50 psi. A six-station manifold is attached to the front of this flask. (Distribution Designs, Inc.)

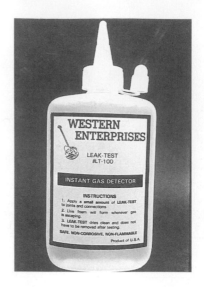

Figure 4-8. The leak test solution is applied to a pressurized joint. The solution will bubble to show location of a leak. (Western Enterprises Co.)

Figure 4-7. This manifold is used for high-pressure gases. A regulator is used to reduce the cylinder pressure to the desired manifold pressure. (Victor Equipment Co.)

Figure 4-9. Station flowmeters of this type operate on manifolds at approximately 50 psi.

gas. Connect the gas to the manifold, set the flow rate on the other end of the pipe to 5 cubic feet per hour (cfh) and purge until analyzer tests show the system is clear of oxygen. When the test shows clear, connect the argon to the line and purge at the same rate for several hours before welding.

If the system requires repair, disassembly, or modification, it must be purged and tested again prior to use.

Gas Regulation

Regulation of shielding gases is accomplished by the use of several types of special equipment. Gases distributed through a manifold require "regulators" like the one shown in Figure 4-7. They reduce the pressure from the cylinder to the desired manifold pressure level, usually in the range of 20 psi to 50 psi. A flowmeter, Figure 4-9, is then used at the welding station. It regulates the flow of gas to the welding gun.

When a cylinder is used at the welding station itself, a ***regulator/flowmeter*** reduces pressure from the cylinder and regulates flow to the gun. See Figures 4-10 and 4-11.

A regulator/flowmeter or a flowmeter with a ball tube should always be installed so that the ball tube is in the vertical position for proper operation. The amount of flow is indicated at the top of the ball, unless otherwise indicated.

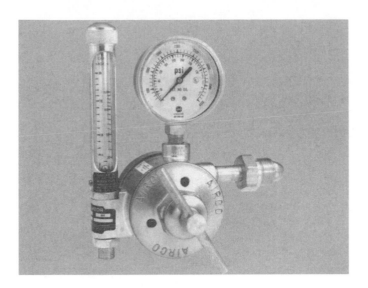

Figure 4-10. Single-cylinder regulator/flowmeters have a gauge that shows cylinder pressure. Gas flow to the gun (in cubic feet per hour) is adjusted by turning the knob and reading the ball position in the vertical tube.

Figure 4-12. Surge check valves like the one below the flowmeter eliminate surging of gas from the nozzle during the start of the operation. They quickly pay for themselves in gas savings. (Weld World Co.)

Regardless of the type of gas supply (cylinder, Dewar flask, manifold), a surge of gas will exit from the gas nozzle when the gas flow valve is opened. This is due to the pressure buildup when the gas is not flowing. This initial gas surge will last for several seconds until the excess pressure is reduced. To eliminate this condition, a specially designed surge check valve may be used. See Figure 4-12.

Gas Mixing

Welding gases can be mixed and stored before they enter the manifold, Figure 4-13. This type of mixer is helpful where large volumes of gases are needed. The gas mixer in Figure 4-14 is used on single-station installations, while the mixer in Figure 4-15 may be employed in single or multiple station installations. Another type of mixer, Figure 4-16, may also be used in single stations. This Y-valve arrangement is often used to achieve gas mixtures with proportions different from the standard ones. The valve is installed on the outlet side of the flowmeter, with the gas metered by two separate gas flowmeters.

To prevent backflow of the gases and improper mixing, a ***backflow check valve***, Figure 4-17, should be

Figure 4-11. A dial gauge is used with a regulator to read gas flow in cubic feet per hour. The desired gas flow is obtained by turning the adjustment knob on the front of the regulator. (Veriflo Corporation)

Figure 4-13. The large tank on the bottom of the mixer serves as a mixing and storage chamber. This assures sufficient volume of mixed gases for large users. (Thermco Instrument Co.)

installed between the flowmeter and the Y valve. Figure 4-18 shows backflow check valves installed on flowmeters.

Mixture and Purity Testing

Gas analyzers, Figure 4-19, test for proper mixes at the welding stations. These instruments can also be used to check for leaks and to establish that pipes or vessels have been adequately purged before welding.

Figure 4-16. A Y valve is a simple means of mixing two gases in proportions different from those provided commercially. (Victor Equipment Co.)

Figure 4-14. Small proportional mixers of this type are used at individual welding stations. (Tescom Corp.)

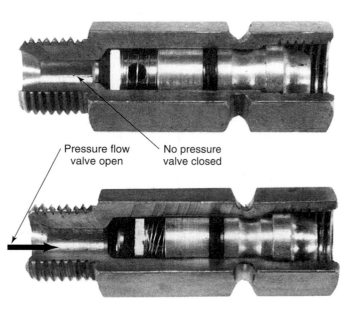

Pressure flow valve open   No pressure valve closed

Figure 4-17. Backflow check valves prevent mixing of gases in the supply line. The upper valve is shown closed. The bottom valve is shown open.

Figure 4-15. The gas mixture from this unit is stated on the label. The gas can be stored or piped directly to the gas manifold for welding use. (Smith's Heating Welding Equipment Co.)

Review Questions

Write your answers on a separate sheet of paper. Do not write in this book.

1. Gases are used in GMAW to _____ the electrode and the molten weld pool from the atmosphere, _____ heat from the electrode to the metal, _____ the arc pattern, _____ in controlling bead contour and penetration, _____ in metal transfer from the electrode, and _____ in the cleaning action of the base material.

2. Two inert gases used in the GMAW process are _____ and _____.

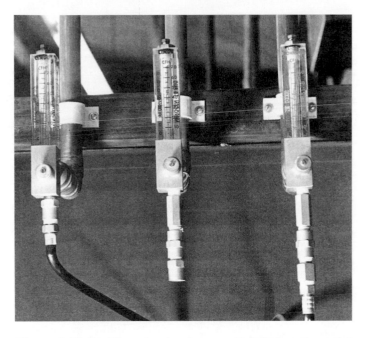

Figure 4-19. Portable analyzers of this type are used to measure the percentages of individual gases in a mixture. (Thermco Instrument Co.)

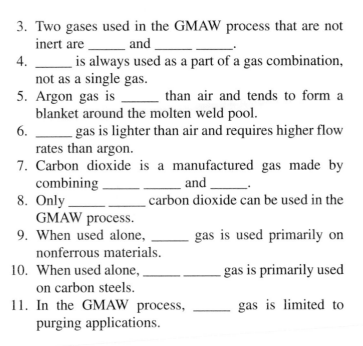

Figure 4-18. Backflow check valves are installed on the outlet side of the flowmeters at center and right.

3. Two gases used in the GMAW process that are not inert are _____ and _____ _____.

4. _____ is always used as a part of a gas combination, not as a single gas.

5. Argon gas is _____ than air and tends to form a blanket around the molten weld pool.

6. _____ gas is lighter than air and requires higher flow rates than argon.

7. Carbon dioxide is a manufactured gas made by combining _____ _____ and _____.

8. Only _____ _____ carbon dioxide can be used in the GMAW process.

9. When used alone, _____ gas is used primarily on nonferrous materials.

10. When used alone, _____ _____ gas is primarily used on carbon steels.

11. In the GMAW process, _____ gas is limited to purging applications.

12. Gas purity is defined as moisture content in _____ _____ _____.

13. Dewar flasks are containers filled with _____ _____.

14. Charges made by a gas supplier for the use of gas cylinders are called _____.

15. Gas cylinders in storage should always have the _____ _____ installed until ready to be used.

16. Cylinders should always be stored and used in the _____ _____ and _____ to a solid object to keep them from falling over.

17. Cylinder gas must always be reduced from the tank pressure to working pressure with a _____ _____.

18. The gas flow used during welding is metered by a _____ to register flow rates in _____ _____ _____ _____.

19. Shielding gases can be checked for purity or gas percentages with a _____ _____.

This DC/CV power source can be used for either short-arc GMAW or FCAW, and is sized for use in operations performing light fabrication or autobody repair work. It can feed wire ranging from 0.023″ to 0.045″ at speeds of up to 750 ipm. (Hobart Welding Products)

Objectives

After studying this chapter, you will be able to:
- Describe how filler materials are made.
- Compare the application of general use, rigid control use, and critical use filler wires.
- Tell how contamination of filler wire is avoided by both the manufacturer and the welder.

To produce acceptable welds, filler materials used in the GMAW process must be of the highest quality. For this reason, manufacturers of welding wire use specialized machines, processes, and inspections. They supply filler wire in roll form to fit various types of wire feeders.

Manufacturing

Material that is to be manufactured into filler wire for the Gas Metal Arc Welding process is selected based on several factors. These may include:
- Chemical composition.
- Mechanical properties.
- Notch toughness values.
- Impurity level limits.

Material selected at the primary mill is *hot-drawn* through reducing dies to specified sizes, then cleaned. Further reduction of the wire is then done cold. *Annealing* (softening) and cleaning operations are performed as the wire is further reduced in size.

Various types of lubricants are used in the dies during the drawing process. The *lubricants* simultaneously reduce die wear and decrease friction between the wire and the dies. Lubricants also carry away heat produced in the dies during the drawing operation.

During drawing, several types of *defects* can occur that will affect the quality of filler materials. These defects include:
- Cracking.
- Seaming.
- Center bursting.
- Oxide formation.
- Overlapping.

Examples are shown in Figure 5-1. Filler materials that exhibit any of these defects should not be used. After drawing, another process cleans the wire of all surface impurities. The material is then processed for shipment to the user.

Specifications

Manufacturers use many quality specifications for the manufacturing, testing, inspecting, and packaging of filler materials. Figure 5-2A list some of the common filler material specifications, while Figure 5-2B list the AWS specifications for GMAW filler materials. Specifications may include the following requirements:
- Scope.
- Classification and usability.
- Manufacturing methods.
- Acceptability tests.

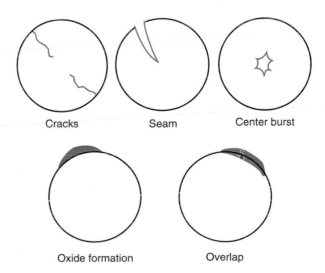

Figure 5-1. Typical defects that can occur in welding wire during manufacturing.

Specifications	Material Types
American Welding Society	All Types
American Society for Testing Materials	All Types
AISI/SAE	Low-alloy Steel
MIL-E-23765	Steel
Aeronautical Material Specifications	All Types

A

Metal	AWS Specification	Classification
Copper Alloys	A5.7	ERCu, ERCuSi-A, ERCuA1-A1
Stainless Steel	A5.9	ER308, ER309, EC409, ER2209
Aluminum Alloys	A5.10	ER1100, ER2319, ER4043, ER5356
Nickel Alloys	A5.14	ERNi-1, ERNiCr-3, ERNiCrMo-3
Carbon Steel Alloys	A5.18	ER70S-2, ER70S-3, ER70S-6, E70C-3M
Magnesium Alloys	A5.19	ERAZ101A, ERAZ61A, EREZ33A
Low-Alloy Steel Alloys	A5.28	ER80S-B2, E80C-B2, ER80S-D2

B

Figure 5-2. Specifications. A—Common wire specifications used by the welding industry to establish quality. B—AWS specifications for GMAW filler materials.

- Chemical composition.
- Usability tests and results.
- Standard sizes.
- Finish and temper.
- Spool and winding requirements.
- Packaging.
- Marking of packages.
- Guarantee.

The user may sometimes add further requirements, such as chemistry, packaging, and weld testing, to these basic specifications. Each additional requirement will add cost to the filler material.

Manufacturers of filler materials *guarantee* only that their product will meet specification. This means they will only replace defective wire. They do not guarantee acceptable results, since they cannot govern the welding application.

Filler wire manufacturers make welding material for three major areas:

- **General use.** This wire will meet specification requirements. However, no record of chemical composition, strength level, or other requirements is submitted to the user when the wire is purchased.
- **Rigid control use.** Fabrication that requires rigid control over the filler material. Wire used under this condition may require a *Certificate of Conformance* with the purchase of the material. This certificate is a statement that the filler material meets all of the requirements of the material specification. An example is shown in Figure 5-3. All of the stock will be identified by heat numbers, lot numbers, or code numbers located on the wire roll package. On some work, the buyer may require these numbers be recorded wherever the material is used in a welding application.
- **Critical use.** Some types of welding operations, such as those involving aircraft, nuclear reactors, and pressure vessels, typically require very close control of the filler material's chemistry. A *Certified Chemical Analysis* report, like the one shown in Figure 5-4, is the actual chemical analysis of the individual heat or lot of material. Testing is done on a machine called a *spectrometer*, Figure 5-5. Records are maintained during fabrication cycles wherever the specific materials are used. In case of a joint failure, either due to the filler material or the base metal, other welded joints in the weldment or system can be evaluated for possible replacement.

Figure 5-3. Certificate of conformance form. (Techalloy Maryland, Inc.)

Figure 5-4. Certified chemical analysis form. (Techalloy Maryland, Inc.)

Figure 5-5. Actual chemical analyses are obtained from a wire sample by use of this machine. (Techalloy Maryland, Inc.)

Filler Material Form

Filler materials used in GMAW are wound on spools or inside large drums, depending on the application. Standard spools are 4″, 8″, or 12″ in diameter, and may be made of plastic, wood, or formed metal wire. The spools are disposable after use. See Figure 5-6.

The wire may be either *level-layer-wound*, as shown in Figure 5-7, or *layer-wound*, as shown in Figure 5-8. The type of winding depends on the material type and the application.

Welding wires are furnished in standard sizes, which include:

- 0.023″ diameter.
- 0.030″ diameter.
- 0.035″ diameter.
- 0.045″ diameter.
- 3/64″ diameter (soft wires).
- 0.062″ diameter.

Different types and sizes of wire have varying quantities of wire on the spool. Nonferrous materials will always have less metal per spool than ferrous materials.

To prevent wire feeding problems during welding, the wire must meet cast and helix requirements. Cast is the diameter of one complete circle of wire from the spool, as it lies on a flat surface. With hard wire, this

Figure 5-6. Various types of wire spools that fit standard wire feeders.

diameter tends to be larger than the diameter on the spool. *Helix* is the maximum height of any point of this circle of wire above the flat surface. Cast and helix dimensions and tolerances are given in Figure 5-9. These two dimensions are critical to the proper feeding of filler material. Improper dimensions will affect the entire wire feed system and may cause the tangled condition called "birdnesting." Other adverse effects of improper wire dimensions are arc outages, severe liner wear, and contact tip wear.

Figure 5-7. The aluminum welding wire on this spool is level-layer-wound.

Identification

Spools or coils of wire are identified by labels or tapes attached to the inside of the spool hub or to the flange. An example of such an identifying label is pictured in Figure 5-10.

Filler Material Packaging

Manufacturers of filler materials use a variety of packaging methods to protect the material during shipment and storage. The spools may be wrapped in a special paper or in plastic to protect the wire. See Figure 5-11. Then, the spools are packed in cardboard cartons or metal cans.

Filler Material Use

Filler materials are easily contaminated by oil, moisture, grease, soot, and salts from the hands. Dirty gloves, rags, or work surfaces will readily contaminate the filler wire whenever they contact it. Such contamination will often cause defects in welds, such as porosity or cracks. Reworking a rejected weld is costly.

Filler material packaging is designed to prevent contamination of the material during shipment. Preventing contamination after the package is opened is the responsibility of the user.

Figure 5-8. The steel welding wire on this spool is layer-wound.

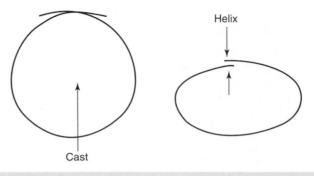

Standards for Cast and Helix							
		Cast				Helix	
Spool Size	Wire Types	Min. in	mm	Max. in	mm	Max. in	mm
4″ (100 mm)	Low-alloy, stainless and nickel alloy	6*	150	9	230	½	13
	Aluminum	†		6	150	1	25
8″ (200 mm)	Low-alloy, stainless and nickel alloy	15*	380	30	760	1	25
	Low-alloy, stainless and nickel alloy	15*	380	30	760	1	25
12″ (300 mm)	Aluminum	†		15	380	1	25

* *Measured on outside strand of full spool*
† *Diameter of wire level from which sample is taken*

Figure 5-9. The terms *cast* and *helix* define the characteristics of any form of continuous wire as it comes from the spool or coil. Cast is the diameter of one complete circle of wire as it lies on a flat surface. Helix is the maximum height of any point of this circle of wire above the flat surface.

Figure 5-10. The label identifies this welding wire as 4043 aluminum, while the letters HQ denote a specially processed high-quality material. The number 5202F07 is a stock number; 707 is the manufacturer's control number. The actual heat number is 0035U. The wire is 3/64″ diameter. The weight of the welding wire on the spool is 14.3 pounds.

Figure 5-11. Spooled wire in this package can be stored indefinitely if the package remains sealed.

The simplest method of avoiding contamination is to keep the filler material clean. Use the following practices:

- Keep the material packaged as long as possible.
- Open packages only when needed.
- Store all unsealed filler materials in a heated cabinet.
- Handle material as little as possible, then only with clean gloves.
- Work in a clean and dry area.
- Remove wire spools from machines whenever machines are to be idle for an extended period of time.

When high quality welds are required, contaminated filler materials cannot be tolerated. Requiring the filler wire manufacturer to clean, inspect, and package the material according to specifications does little good if the material is contaminated by improper handling prior to use. Figure 5-12 shows what can happen when a roll of steel wire is left exposed to the elements. The wire is no longer usable and must be discarded.

Figure 5-12. If wire is allowed to rust on the spool, it will result in weld defects that could cause the weld to fail.

Figure 5-13. Allowing wire to tangle on the spool is a costly mistake that can easily be prevented.

Wire must be properly handled before being stored. Figure 5-13 illustrates what can happen when a roll of wire is removed from a machine and the loose end of the wire is released before being secured on the reel. The wire overlaps and tangles to the point where it must be discarded. To prevent this, tie the end of the wire to the spool after removing the spool from the machine, as shown in Figure 5-14.

Filler Material Selection

One of the most important factors to consider in GMAW is selecting the correct filler wire. The chosen wire, in combination with the shielding gas, will produce the deposit chemistry that determines the final physical and mechanical properties of the weld. These basic factors influence the choice of filler wire:

- Base metal chemical composition.
- Base metal mechanical properties.
- Type of shielding gas.
- Weld joint design.
- Service use of the weldment.

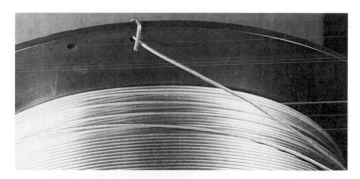

Figure 5-14. Each time the wire is removed from the machine, secure the end to prevent tangling. Loop the wire end through one of the holes provided in the outer edge of the reel.

The actual process of selecting a filler wire for a particular welding operation is described in the chapters relating to specific materials.

Review Questions

Write your answers on a separate sheet of paper. Do not write in this book.

1. What are the four factors that are used to select welding wire at the mill for drawing into various diameters?
2. As wire is drawn through a die to reduce its diameter, it becomes hard and must be _____ to make it soft.
3. List the five types of defects produced during the drawing operation that can make the wire unusable.
4. One of the major specifications used for the purchase of welding wire is made by the _____ _____ _____.
5. _____ _____ wire is obtained for welding when no specifications have been defined by the manufacturer of the welded product.
6. What document is used when manufacturing specifications require a record of wire quality?
7. A report of the actual chemical composition of the material is stated on a _____ _____ _____.
8. Welding wire is wound on spools of _____", _____", or _____" diameter.
9. Wire manufactured to a fractional-inch size dimension is considered to be a _____ wire.
10. Major problems may be encountered in feeding spooled wire during the welding operation if the correct _____ and _____ requirements are not met.
11. Proper _____ and _____ of welding wire are vital to preventing contamination of the filler wire.
12. List the five basic factors that influence the choice of filler material to be used in welding.

CHAPTER 6

Weld Joints and Weld Types

Objectives

After studying this chapter, you will be able to:
- Identify different types of welds used with various joints.
- Read and draw common welding symbols.
- Discuss advantages and disadvantages of different weld joints.
- List factors involved with joint design.

Joint Types

The American Welding Society defines a joint as "the manner in which materials fit together." As shown in Figure 6-1, there are five basic types of weld joints:
- Butt joint.
- T-joint.
- Lap joint.
- Corner joint.
- Edge joint.

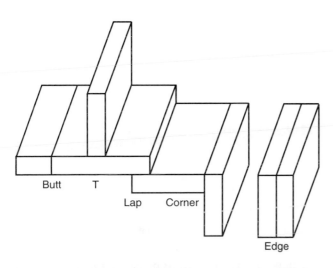

Figure 6-1. The five basic types of joints used in welding.

Butt T Lap Corner Edge

Joint Preparation

Weld joints may be initially prepared in a number of ways. These include:
- Shearing.
- Casting.
- Forging.
- Machining.
- Stamping.
- Filing.
- Routing.
- Oxyacetylene cutting (thermal cutting process).
- Plasma arc cutting (thermal cutting process).
- Grinding.

Final preparation of the joint prior to welding will be covered in the chapters that detail the welding of specific materials.

Weld Types

There are various types of welds that can be made in each of the basic joints. They include:

Butt joint, Figure 6-2.
- Square-groove butt weld.
- Bevel-groove butt weld.
- V-groove butt weld.
- J-groove butt weld.
- U-groove butt weld.
- Flare-V-groove butt weld.
- Flare-bevel-groove butt weld.

T-joint, Figure 6-3.
- Fillet weld.
- Plug weld.
- Slot weld.
- Bevel-groove weld.
- J-groove weld.
- Flare-bevel-groove weld.
- Melt-through weld.

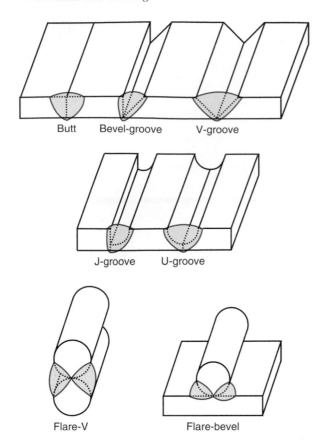

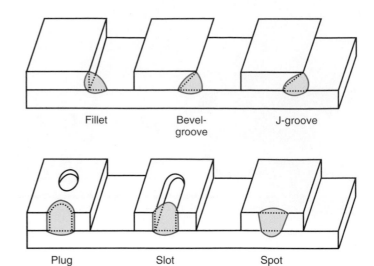

Figure 6-4. Types of welds that may be made with a basic lap joint.

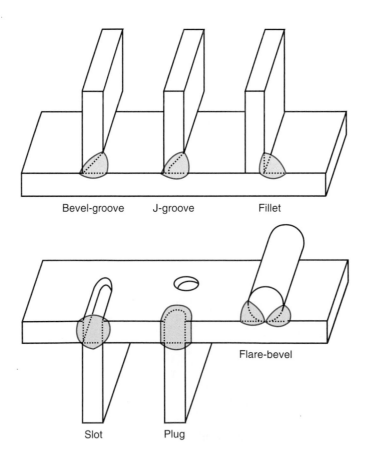

Figure 6-2. Types of welds that may be made with a basic butt joint.

Figure 6-3. Types of welds that may be made with a basic T-joint.

Lap joint, Figure 6-4.
- Fillet weld.
- Plug weld.
- Slot weld.
- Spot weld.
- Bevel-groove weld.
- J-groove weld.
- Flare-bevel-groove weld.

Corner joint, Figure 6-5.
- Fillet weld.
- Spot weld.
- Square-groove weld or butt weld.
- V-groove weld.
- Bevel-groove weld.
- U-groove weld.
- J-groove weld.
- Flare-V-groove weld.
- Edge weld.
- Corner-flange weld.

Edge joint, Figure 6-6.
- Square-groove weld or butt weld.
- Bevel-groove weld.
- V-groove weld.
- J-groove weld.
- U-groove weld.
- Edge-flange weld.
- Corner-flange weld.

Double Welds

In some cases, a weld cannot be made from only one side of the joint. When a weld must be made from both sides, it is known as a ***double weld***. Figure 6-7 shows common applications of double welds in basic joint designs.

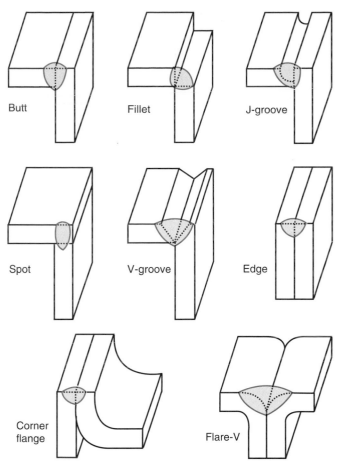

Figure 6-5. Types of welds that may be made with a basic corner joint.

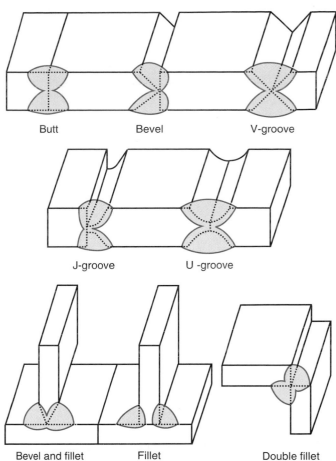

Figure 6-7. Applications of double welds.

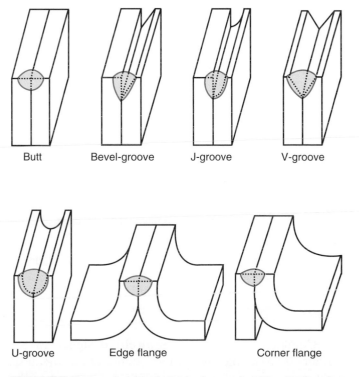

Figure 6-6. Types of welds that may be made with a basic edge joint.

Weldment Configurations

The basic joint often is changed to assist in a component's assembly. A weld joint might be modified to gain access to the weld joint or to change a weld's metallurgical properties. Some common weldment configuration designs are described here. ***Joggle-type joints*** are used in cylinder and head assemblies where backing bars or tooling cannot be used. See Figure 6-8. Another application of joggle joints is in the repair of unibody automobiles where skin panels are placed together and welded. A built-in backing bar is used when enough material is available for machining the required backing or when tooling cannot be inserted (as in some tubular applications). An example in which tubing is being joined to heavy wall tube is shown in Figure 6-9. Pipe joints often use special backing rings or are machined to fit specially designed mated parts. Types of backing rings are shown in Figure 6-10. Figure 6-11 shows a fabricated backing bar. These bars must fit tightly or problems will be encountered in heat flow and penetration. Weld joints specially designed for controlled penetration are used where excessive weld penetration would cause a problem with assembly or liquid flow. This type of joint is shown in Figure 6-12.

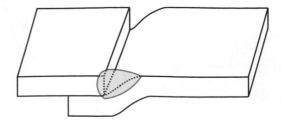

Figure 6-8. Joggle-type joint.

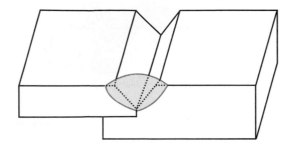

Figure 6-9. Tubular butt joint with a built-in backing bar.

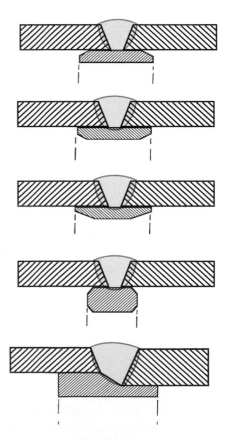

Figure 6-10. Various types of backing rings for pipe joints.

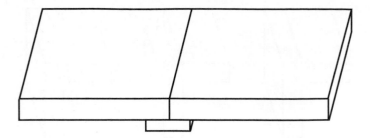

Figure 6-11. Plate butt weld with a fabricated backing bar.

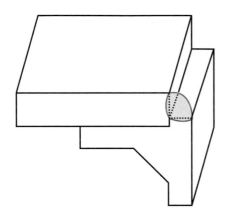

Figure 6-12. Controlled weld penetration joint.

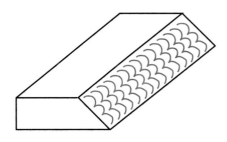

Figure 6-13. Buttered weld joint face.

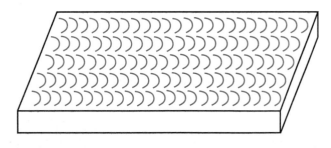

Figure 6-14. Overlaid welds, called surfacing or cladding, protect the base metal from wear or contamination.

Welding Terms and Symbols

A series of bead welds overlaid on the face of a joint is called *buttering,* Figure 6-13. Buttered welds are often used to join dissimilar metals. A series of overlaid welds on the surface of a part to protect the base material is called *surfacing* or *cladding*. Refer to Figure 6-14.

Communication from the weld designer to the welder is essential to proper completion of most weldments. Some of the common terms used to describe parts of the weld joint are found in Figure 6-15. Other terms used to describe welds are given in Figure 6-16. The *AWS welding symbol* shown in Figure 6-17 was developed as a standard by the American Welding Society. This symbol is used on drawings to indicate the type of joint, placement, and the type of weld to be made.

The symbol may also include other information, such as finish and contour of the completed weld.

It is important to study and understand each part of the welding symbol. Figure 6-18 is a table showing basic *weld symbols* that are used with the AWS welding

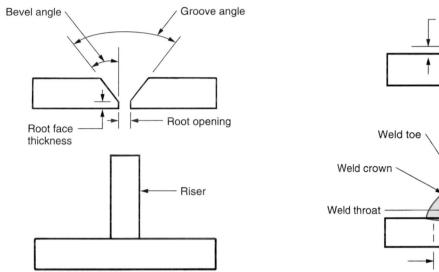

Figure 6-15. Weld joint terms.

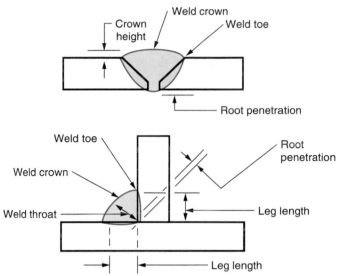

Figure 6-16. Weld and weld area terms.

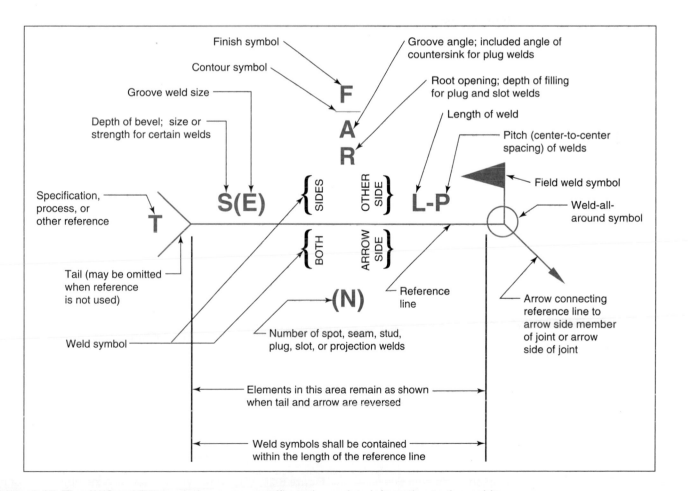

Figure 6-17. The AWS welding symbol conveys specific and complete information to the welder. (Printed with permission of the American Welding Society)

Groove							
Square	Scarf	V	Bevel	U	J	Flare-V	Flare-bevel
(symbols)	(symbols)	(symbols)	(symbols)	(symbols)	(symbols)	(symbols)	(symbols)
Fillet	Plug or slot	Stud	Spot or projection	Seam	Back or backing	Surfacing	Edge
(symbols)	(symbols)	(symbols)	(symbols)	(symbols)	(symbols)	(symbols)	(symbols)

Figure 6-18. Basic weld symbols. (Printed with permission of the American Welding Society)

symbol to direct the welder in producing the proper weld joint. The arrow of the welding symbol indicates the point at which the weld is to be made. The line connecting the arrow to the *reference line* is always at an angle. Whenever the basic weld symbol is placed below the reference line, as shown in Figure 6-19, the weld is made on the side where the arrow points (referred to as the *arrow side*). Whenever the basic symbol is placed above the reference line, the weld is to be made on the *other side* of the joint, as shown in Figure 6-20. By placing dimensions on the symbol and drawings, the exact size of the weld may be indicated. Study the

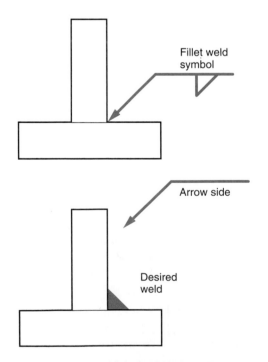

Figure 6-19. Fillet weld symbol shown on the bottom side of the reference line indicates that the weld is located where the arrow points.

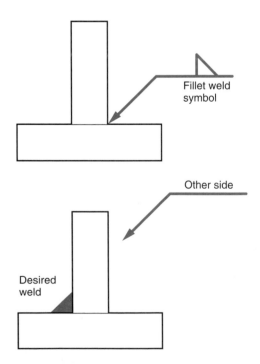

Figure 6-20. Fillet weld symbol shown on the upper side of the reference line indicates that the weld is located on the opposite side of the joint.

examples of typical weld symbols and weldments shown in Figure 6-21.

The complete weld symbol gives the welder instructions on how to prepare the base metal, the welding process to use, and the finish for the completed weld. Through careful use of these symbols, the weld designer can convey all the information needed to complete a weldment.

Classes are offered that provide advanced study in the area of print reading for welders. By taking such classes, the welder can improve his or her ability to read and interpret welding drawings. Studying texts on print reading is another method of gaining ability to read prints.

Weld Positions

For a welder, it is important to be able to weld in different positions. The American Welding Society has defined the positions of welding to include:

- Flat.
- Horizontal.
- Vertical.
- Overhead.

Figure 6-22 demonstrates the four positions for fillet welds, grooved butt welds, and pipe welds. While practicing welding in these positions, you should note how *gravity* affects the molten weld pools. In addition to this, *heat distribution* also varies with each position. These factors make the skills needed for each position distinct. Practice is required to produce good welds in all positions.

Design Considerations

Design of the weld type and weld joint to be used is of prime importance if the weldment is to do the intended job. The weld should be made at reasonable cost. Several factors concerning the weld design must be considered:

- Material type and condition (annealed, hardened, tempered).
- Service conditions (pressure, chemical, vibration, shock, wear).
- Physical and mechanical properties of the completed weld and heat-affected zone.
- Preparation and welding cost.
- Assembly configuration and weld access.
- Equipment and tooling.

Butt Joints and Welds

Butt joints are used where high strength is required. They are reliable and can withstand stress better than any other type of weld joint. To achieve full stress value, the weld must have 100 percent penetration through the joint. This can be done by welding completely through from one side. The alternative is working from both sides, with the welds joining in the center.

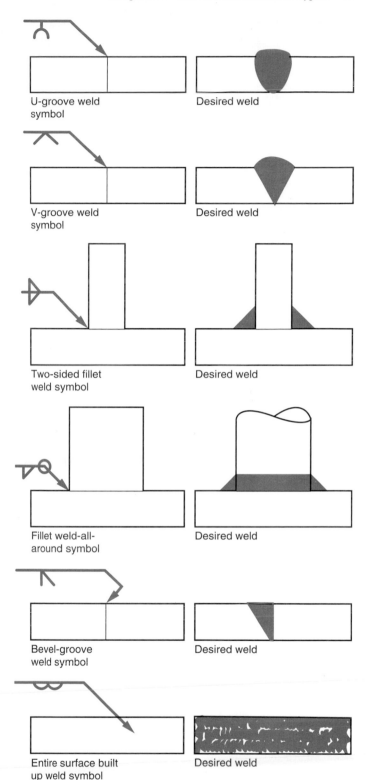

U-groove weld symbol — Desired weld

V-groove weld symbol — Desired weld

Two-sided fillet weld symbol — Desired weld

Fillet weld-all-around symbol — Desired weld

Bevel-groove weld symbol — Desired weld

Entire surface built up weld symbol — Desired weld

Figure 6-21. Typical weld symbols and weld applications.

Thinner-gauge metals are more difficult to fit up for welding. Thin metals also require more costly tooling to maintain the proper joint configuration. ***Tack welding*** may be used as a method of holding the components during assembly. However, tack welds present many problems:

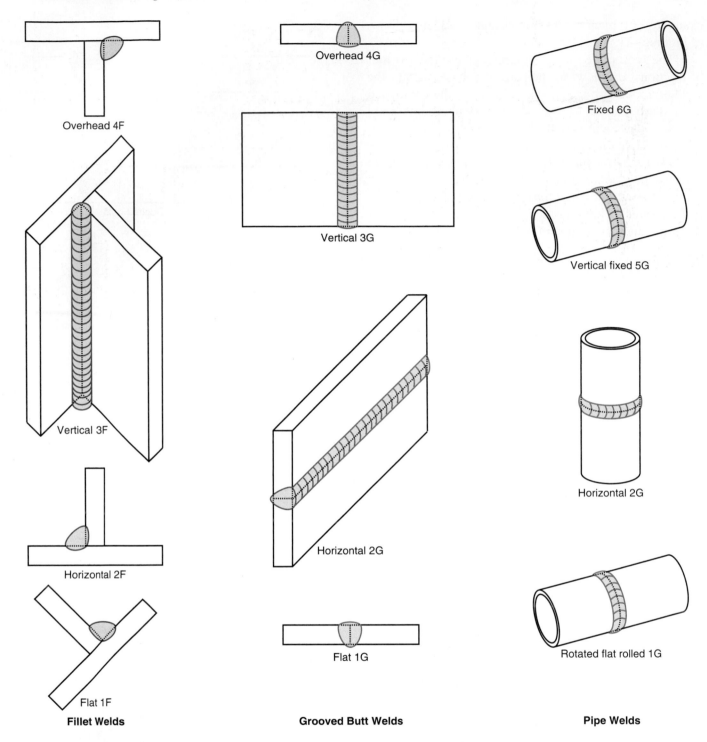

Figure 6-22. American Welding Society definitions of welding positions.

- They conflict with penetration of the final weld into the weld joint.
- They add to the crown dimension (height).
- They often crack during welding due to the heat and expansion of the joint.

Expansion of the base metal during welding often will cause a condition known as *mismatch*, Figure 6-23. When mismatch occurs, the weld generally will not penetrate completely through the joint. Many specifications limit highly stressed butt joints to a 10 percent maximum mismatch of the joint thickness.

Whenever possible, butt joints should mate at the bottom, Figure 6-24. Joints of unequal thickness should be tapered in the weld area to prevent incomplete or inadequate fusion. This is shown in Figure 6-25. When this cannot be done, the heavier piece may be tapered on the upper part of the joint as well.

Weld shrinkage. Butt welds always shrink across the joint (transversely) during welding. For this reason, a shrinkage allowance must be made if the "after welding" overall dimensions have a small tolerance. Butt welds in pipe, tubing, and cylinders also shrink on the diameter of

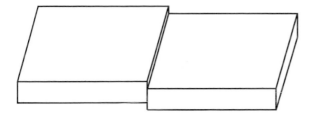

Figure 6-23. Welds made on mismatched joints often will fail below the rated load when placed in stress conditions.

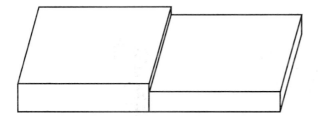

Figure 6-24. Mating the joint at the bottom equalizes the load during stress when the weld is made from the top and penetrates completely through the joint.

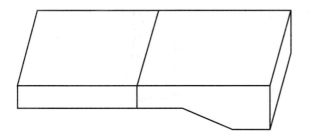

Figure 6-25. Joints of unequal thickness absorb different amounts of heat and expand at different ratios. Equalize the heat flow by tapering the heavier material to the thickness of the thinner material.

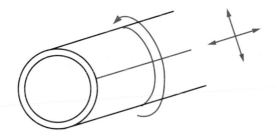

Figure 6-26 Butt welds shrink during welding in both transverse and circumferential directions.

the material. This shrinkage is shown in Figure 6-26. In areas where these dimensions must be maintained, a shrinkage test must be done to establish the amount of shrinkage. Figure 6-27 shows how such a test is made. Heavier materials will shrink more than thinner materials. Double-groove welds will shrink less than single-groove welds. This is because less welding is involved and less filler material is used.

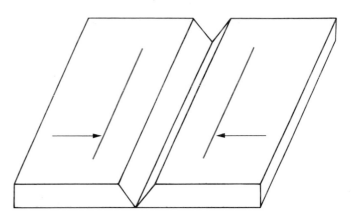

Figure 6-27. Weld joint shrinkage can be determined in four steps. 1. Tack weld the test joint together. 2. Scribe parallel lines, as shown, on approximately 2" centers. Record this dimension. 3. Weld joint with test weld procedure. 4. Measure linear distance and compare with original dimension.

T-Joints and Welds

Various T-joint designs are used to join parts at an angle to each other. Depending on the intended use of the weldment, the joint may be made with a single fillet, double fillet, or a groove and fillet weld combination. Figure 6-28 shows these designs.

Fillet welds are made to specific sizes that are determined by the allowable design load. They are measured as shown in Figure 6-29. Where design loads are not known, a "rule of thumb" may be used for determining the fillet size. In these cases, the fillet weld leg lengths must equal the thickness of the thinner material.

The main problem in making fillet welds is lack of penetration at the joint intersection. To prevent this condition, always make stringer beads at the intersection. Weave beads do not provide the desired penetration on fillet welds.

Lap Joints and Welds

Lap joints may be either single fillet, double fillet, plug slot, or spot-welded. They require very little joint preparation. They are generally used in static load applications or in the repair of unibody automobiles. Where corrosive liquids are involved, both edges of the joint must be welded. See Figure 6-30. One of the major problems with lap joint design is shown in Figure 6-31. Where the component parts are not in close contact, a ***bridging fillet weld*** must then be made. This leads to incomplete fusion at the root of the weld and oversize fillet weld dimensions. When using this type of design in sheet or plate material, clamps or tooling must be used to maintain adequate contact of the material at the weld joint.

An ***interference fit*** eliminates this problem in assembly of cylindrical parts, Figure 6-32. The inside diameter of the outer part is made several thousandths of an inch smaller than the outside diameter of the inner

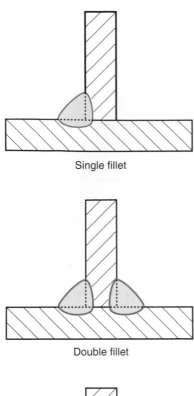

Single fillet

Double fillet

Bevel and fillet

Figure 6-28. Various types of T-joints and welds.

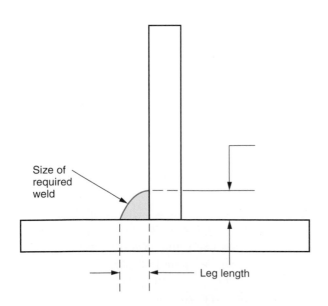

Size of required weld

Leg length

Figure 6-29. Fillet weld leg lengths from the root of the joint should be equal. Unequal leg length, unless otherwise specified, will not carry the designed load and may fail under stress.

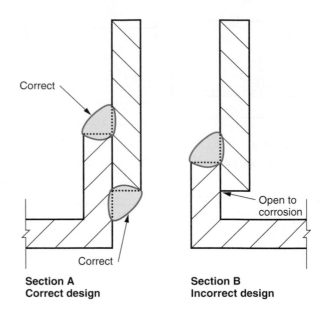

Correct

Open to corrosion

Correct

**Section A
Correct design**

**Section B
Incorrect design**

Figure 6-30. Corrosive liquids must not be allowed to enter the penetration side of the weld joint. In Section A, the back of the weld is closed to corrosion. In Section B, the back of the weld is open to corrosion.

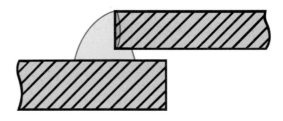

Figure 6-31. Lap joint problem areas that result from improper fit-up.

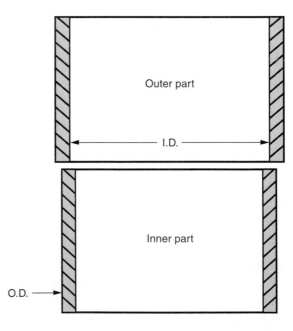

Outer part

I.D.

Inner part

O.D.

Figure 6-32. The diameters of the parts to be assembled with an interference fit may be found by using a "pi" tape around the inner and outer cylinder components. The tape measures in thousandths of an inch and full inches.

part. Before assembly, the outer part is heated until it expands enough to slide over the inner part. As the part cools, it shrinks and locks the two pieces together.

Corner Joints and Welds

Corner joints are similar to T-joints, since they consist of sheets or plates mating at an angle to one another. They are usually used in conjunction with groove welds and fillet welds. Some of the many different corner weld designs are shown in Figure 6-33. When using thinner gauges of metal, it may be difficult to assemble component parts without proper tooling. Tack welding and welding often will cause distortion and buckling of thinner materials. For the most part, use of corner joints should be limited to heavier materials in structural assemblies.

Edge Joints and Welds

Edge welds are used where the edges of two sheets or plates are adjacent and are in approximately parallel planes at the point of welding. Figure 6-34 shows several types of edge weld designs. These designs are common only in structural use. Since the weld does not penetrate completely through the joint thickness, it should not be used in stress or pressure applications.

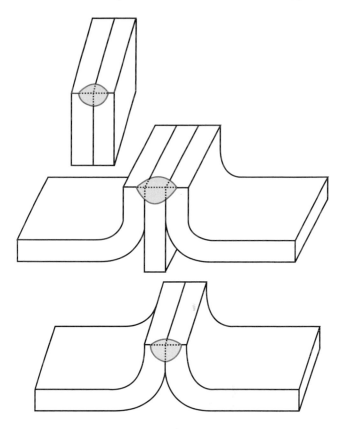

Figure 6-34. Common edge weld joint designs that may be used in fabrication.

Special Designs and Procedures

Special designs are often used in the fabrication of a weldment where:

- Welds cannot be thermally treated after welding because of configuration or size. Material is added to the joint thickness, as shown in Figure 6-35, and the weld is made restricting the heat flow into the thinner metal. This type of joint will achieve the full mechanical values of the base material.
- Joining of dissimilar materials can be done by buttering the face of one material to match the other, as shown in Figure 6-36.

Special procedures and tooling may be used to provide a preheating, interpass, and postheating operation to control grain size. **Preheating** is generally used to slow down the cooling rate of the weld to prevent cracking. *Interpass temperature* is the Min/Max temperature at which a weld can be made on a multi-pass weld. Individual chapters regarding the welding of various metals will define the requirements for preheating, interpass, and postheating temperatures. These temperatures may be checked by the use of special crayons, paints, or pellets like those in Figure 6-37.

Special procedures, tooling, and chill bars may be used to localize and remove welding heat during the welding application. Figure 6-38 shows an application of tooling used to remove heat from the part.

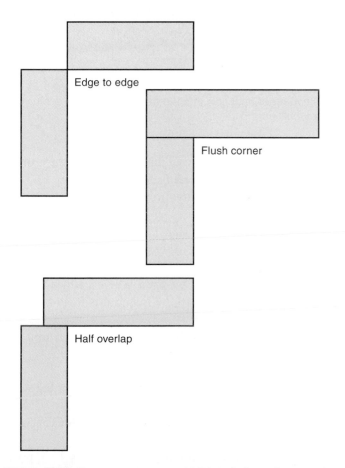

Figure 6-33. Common corner weld joint designs that may be used in fabrication.

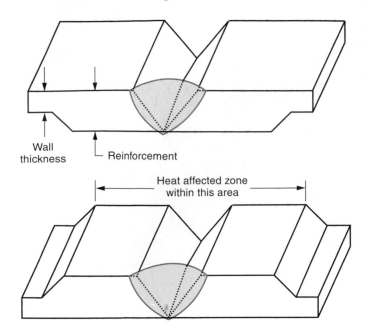

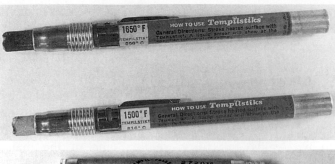

Figure 6-37. Temperature crayons and pellets may be used to determine preheating temperatures prior to welding.

Figure 6-35. The joint thickness and the filler material tensile strength is equivalent to the strength of the base material in this design.

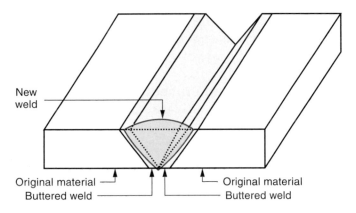

Figure 6-36. Buttering techniques are commonly used to adapt dissimilar metals for welding.

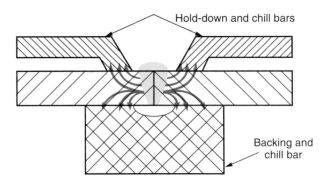

Figure 6-38. Tooling and bars used to remove heat from the weld and resist heat flow into the base material are sometimes called "chill bars."

Review Questions

Write your answers on a separate sheet of paper. Do not write in this book.

1. What are the five basic types of joints?
2. Welds made from both sides of the joint are called _____ welds.
3. _____ welds are often used where dissimilar materials are joined.
4. Weld material placed on the surface of a joint for protection of the base material is called _____ or _____.
5. The oxyacetylene and the plasma arc processes are _____ cutting processes.
6. Where high-strength welds are required, a _____ joint weld with 100 percent penetration is used.
7. Unless tack welds are made properly, they often _____ during the welding operation.
8. Fillet welds are always measured by the length of the fillet _____.
9. Where fillet weld dimensions are not specified, the size of the weld should equal the thickness of the _____ material.
10. The condition of a butt joint where neither the top nor the bottom edge of the material is flush is called _____.
11. _____ joints always shrink across the weld joint.
12. The main problem in making fillet welds is _____ _____ _____ into the root of the joint.
13. Edge-type joints without full penetration of the weld should not be used in _____ _____.
14. _____ bars are used during welding to localize and remove heat from the weld area.
15. The American Welding Society defines four positions of welding: _____, _____, _____, and _____.

CHAPTER 7

Welding Procedures and Techniques

Objectives

After studying this chapter, you will be able to:
- Adjust welding current and voltage parameters to produce desired qualities in welds.
- Vary electrode extension, travel speed, direction of travel, gun angle, and bead pattern to achieve desired weld qualities.
- Determine values for a welding schedule, based on results from test welds.

Basic Operation

The GMAW process is considered to be either a semiautomatic or a fully automatic operation. In the *fully automatic* operation, all the parameters and variables are controlled by the machine; operator skill is not essential. In the *semiautomatic* operation, however, the hands-on skill of the welder becomes important in the final quality of the weld. The welder must be able to set up the machine properly and operate the gun in such a manner that the weld will meet the weld quality requirements of the fabrication specification.

Welding Variables and Parameters

Welding variables are determined after the material type and thickness, and mode of welding have been selected. These parameters include:

Welding current/amperage (wire speed). With a constant voltage machine, the welding current is established by the *wire speed*. Increasing or decreasing the wire speed will result in an increase or decrease in welding current.

Wire speed is always established in *inches per minute (ipm)*. This parameter may be determined by using the following sequence:
1. Turn the machine on and run the wire to the end of the gas nozzle or the contact tip.

2. Depress the arc start switch (this starts wire feed) and run for 10 seconds.
3. Measure wire length from the end of the contact tip or the gas nozzle, as shown in Figure 7-1.
4. Multiply wire length by six (10 seconds = 1/6 of a minute). The resulting number will be the wire speed in inches per minute.

Keep the end of the wire away from the ground. If contact is made, an arc will occur.

Welding voltage. The *welding voltage* is the actual arc gap established between the end of the electrode and the workpiece. This parameter is set on the welding machine. The machine automatically changes the amperage output to maintain this preset arc voltage.

Reference charts listing the initial welding parameters for steel, stainless steel, and aluminum are located in the reference section.

Electrode extension. *Electrode extension* is the distance from the contact tip to the end of the electrode. Often, the term *stickout* is used to define the length of unmelted electrode. The dimensions used for these terms are shown in Figure 7-2. The welder controls electrode extension through the handling of the gun while welding. Changing the extension from the originally established dimension has a marked effect on the melting of the electrode. Increasing the extension will cause more

Figure 7-1. To establish wire speed, carefully measure the length of wire fed in 10 seconds. Wire speed directly affects the welding current.

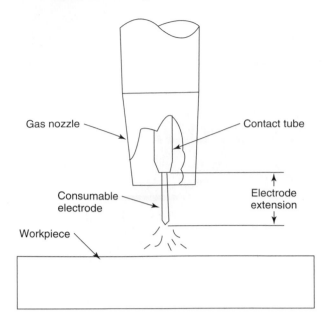

Figure 7-2. Electrode extension. The actual electrode extension is from the end of the contact tip to the end of the electrode. The proper term is "electrical stickout" and is defined on weld procedures with the letters ESO.

resistance heating of the wire and a lower welding current flow. This is useful when welding on gapped joints to prevent burnthrough. Conversely, decreasing the extension makes less wire available for resistance heating. This increases current flow for greater penetration. Figure 7-3 shows this relationship.

Travel speed. The travel speed of the welding gun must be regulated to produce a good weld. In machine welding, the speed is set in inches per minute (ipm). In manual welding, the welder controls this variable. The welder must be able to adapt to changing shapes, improper fit-up, gaps, and other variables as the weld continues. This is where the skill of the welder is very important to the quality of the weld.

Direction of travel. The GMAW operation can be done with the welding gun pointing back at the weld and the weld progressing in the opposite direction. This is

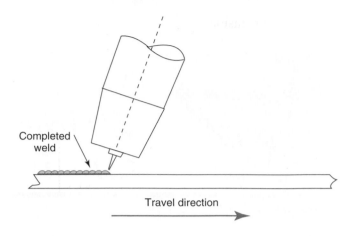

Figure 7-4. Backhand or pull welding technique. This technique requires more skill on the part of the welder, since the weld joint is difficult to see because of the position of the gas nozzle.

called ***backhand welding*** or ***pull welding*** and is shown in Figure 7-4. Some of the advantages of this method include:

- More stable arc.
- Less spatter.
- Deeper penetration.

The other method of gun manipulation is to point the gun forward in the direction of travel. This method is called ***forehand welding*** or ***push welding***. This is shown in Figure 7-5. In comparison to backhand welding, this method results in:

- More spatter.
- Less penetration.
- Better visibility of the weld seam for the welder.
- Good cleaning action of the base metal weld area when welding aluminum.

Gun angle. Gun angles are defined in either ***longitudinal*** (along the weld) or ***transverse*** (across the

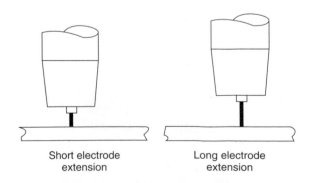

Figure 7-3. Electrode extension variations. Use these variations from the welding procedure dimensions to increase or decrease welding heat into the molten weld pool.

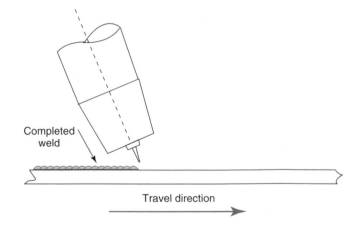

Figure 7-5. Forehand or push welding technique. The weld joint is directly in front of the electrode, so this technique is easy to use. When using the forehand technique, remember that there will be less penetration into the joint and that spatter will increase.

weld) angle dimensions. The longitudinal angle may also be referred to as the travel angle, and the transverse angle may be referred to as the work angle. These are found in Figure 7-6. Gun angle affects weld penetration, bead form, and final weld bead appearance. The angles shown for the various welds are to be used as a starting point for your weld. As the weld progresses, changes in gun angle may be required to maintain good weld pool control and weld bead shape.

Figure 7-7 shows the transverse gun angles and bead placement for flat groove welds. Figure 7-8 shows the transverse gun angles and bead placement for horizontal groove welds.

Figure 7-9 shows the transverse gun angles and bead placement for horizontal fillet welds. Figure 7-10 shows the gun angles for groove and fillet welding *uphill* (commonly called "vertical-up").

Figure 7-11 shows the gun angles for groove and fillet welding *downhill* (commonly called "vertical-down").

Weld bead patterns. Two types of patterns are used for depositing metal. They include:

- Stringer bead pattern. In the *stringer bead* pattern, travel is along the joint with very little side-to-side motion. It may be made with a small zig-zag motion or in a small circular motion. These motions are pictured in Figure 7-12.

- Weave bead pattern. The *weave bead* pattern may also be called a wash bead pattern or an oscillation bead pattern. The weld is wider than a stringer bead and requires a dwell (wait) of the gun at the end of each weave to fill the metal into the weld without undercut. This is shown in Figure 7-13.

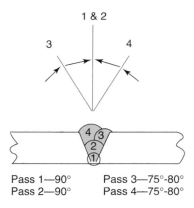

Pass 1—90°   Pass 3—75°-80°
Pass 2—90°   Pass 4—75°-80°

Figure 7-7. Flat groove-weld gun angles and bead placement. Longitudinal angles are 10°-15° from the vertical position. The angles shown are transverse angles and are only for initial welding set-up. Each variation in groove angle or thickness dimension will affect the required gun angle. Remember, the molten weld pool must flow into the previous pass and the side walls of the joint. Develop your welding skill so that you can properly fill the groove joint with a sound weld.

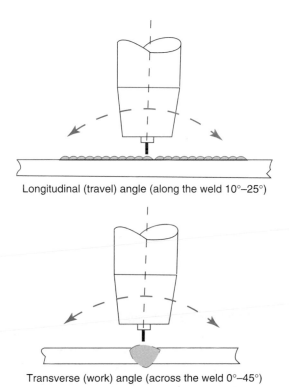

Longitudinal (travel) angle (along the weld 10°–25°)

Transverse (work) angle (across the weld 0°–45°)

Figure 7-6. Gun angles along the joint are referred to as longitudinal angles, or travel angles. Longitudinal angles generally range between 10° and 25° degrees off perpendicular. Gun angles across the joint are called transverse angles, or work angles, and range from 0° to 45° depending on the welding position and joint type.

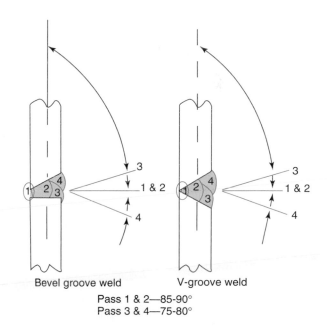

Bevel groove weld   V-groove weld
Pass 1 & 2—85-90°
Pass 3 & 4—75-80°

Figure 7-8. Horizontal groove-weld gun angles and bead placement. Longitudinal angles are 10°-15° from the vertical position. The joint on the left is a bevel-groove joint with the gun transverse angles shown for initial setup only. Remember to make the weld beads smaller when welding out of position and always make stringer beads. (Do not make weave beads.) Fill each layer from the bottom upward. The joint on the right is a V-groove joint with the gun transverse angles shown for initial setup only. The same rules apply as in a bevel-groove joint.

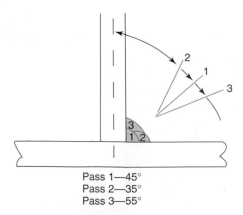

Pass 1—45°
Pass 2—35°
Pass 3—55°

Figure 7-9. Horizontal fillet weld gun angles and bead placement. Longitudinal angles are 10°–15° from the vertical position. The angles shown are transverse angles and are only for initial welding setup. For a single pass weld, use the 45° angle and use the other angles for a multiple pass weld. On a multiple pass weld, always make the second pass on the bottom. The completed weld should have a slightly convex crown and equal length legs. The size of the fillet should be equal to the thickness of the thinnest material used in the joint.

Figure 7-10. Vertical groove and fillet weld gun placement when welding uphill (vertical up). The groove weld is shown on the left. The fillet weld is shown on the right. The bead placement for the fillet weld is the same as for flat and horizontal welds. Weave beads may be used after the root pass is made and if the joint access is sufficient to accept the additional metal.

Figure 7-11. Vertical groove and fillet weld gun placement when welding downhill (vertical down). The groove weld is shown on the left, the fillet weld on the right. The bead placement for fillet welds is the same as for flat and horizontal welds. Make stringer beads and keep the electrode on the front edge of the molten weld pool.

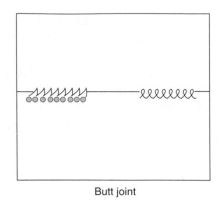

Butt joint

Figure 7-12. Stringer weld bead patterns. The bead on the left is called a backstep or zig-zag technique and requires a dwell point on each swing backward. These points are marked with a colored dot. The circular bead technique shown on the right does not have any dwell points. As you practice these various techniques, vary your forward speed as necessary to maintain an even bead contour and shape.

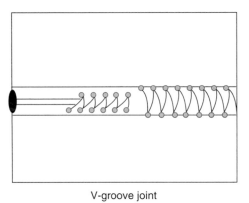

V-groove joint

Figure 7-13. Weave bead patterns. The weave bead pattern is prone to undercut at the outer edges of the weld and requires a hold or dwell point as indicated by the colored dot.

Welding Schedules

The actual welding of component parts requires the combining of many parameters and variables. It often would be difficult to remember all the materials, wire sizes, types of gases, and techniques used to make the weld. Therefore, a listing of all these items can be placed on a form called a *weld schedule*, Figure 7-14.

To determine the actual values placed on the form, select an initial setting from the charts shown in the reference section for the type and thickness of material involved. The next step is to clean the material and make test welds. During the test weld period, adjust the machine and weld techniques to produce the desired weld. Listen to the sound of the arc as you are welding. When short-arc welding, the molten wire should leave the end of the wire evenly, without excessive noise and popping as it contacts the weld pool. A buzzing sound indicates the arc voltage and wire feed are properly set.

GAS METAL ARC WELDING (GMAW) SCHEDULE

WELD TYPE _____ POSITION _____

BASE METAL TYPE _____ THICKNESS _____

WIRE TYPE _____ DIAMETER _____ SPECIFICATION _____

WIRE SPEED (IPM) _____ AMPERAGE _____ WIRE STICKOUT _____

MODE _____ VOLTAGE _____ INDUCTANCE _____

GAS TYPE _____ % GAS TYPE _____ % CFH _____

NOZZLE TYPE _____ NOZZLE DIAMETER _____

TRAVEL FOREHAND _____ TRAVEL BACKHAND _____

UPHILL _____ DOWNHILL _____ PRE-WELD CLEANING _____

INTERPASS CLEANING _____

PREHEAT_____ INTERPASS TEMP _____ POSTHEAT _____

NOTES AND SPECIAL INSTRUCTIONS:

Figure 7-14. Weld schedule. Forms like this one are used by industry to record data for ready reference for each welding job.

WELDING VARIABLES TO CHANGE	DESIRED CHANGES							
	PENETRA-TION		DEPOSITION RATE		BEAD SIZE		BEAD WIDTH	
	↑	↓	↑	↓	↑	↓	↑	↓
CURRENT & WIRE FEED SPEED	↑	↓	↑	↓	↑	↓	*	*
VOLTAGE	+	+	*	*	*	*	↑	↓
TRAVEL SPEED	+	+	*	*	↓	↑	↑	↓
STICKOUT	↓	↑	↑	↓	↑	↓	↓	↑
WIRE DIAM.	↓	↑	↓	↑	*	*	*	*
SHIELD GAS % CO_2	↑	↓	*	*	*	*	↑	↓
TORCH ANGLE	BACK-HAND TO 25°	FORE-HAND	*	*	*	*	BACK-HAND	FORE-HAND

* NO EFFECT   ↑ INCREASE   ↓ DECREASE
+ LITTLE EFFECT

Figure 7-15. Adjustments in welding parameters and techniques.

When spray-arc welding, the sound will be a hissing that should be continuous without popping and interruption.

Figure 7-15 shows various types of adjustments that can be made in procedures and techniques. Chapter 11 of this text shows many of the common defects that occur during welding, and steps that can be taken to correct these conditions.

Each welder develops individual techniques for many types of welds. For this reason, do not expect that all the welds will look alike. Machine welds *can* be duplicated and the tolerance for the parameters can be smaller. In the case of welds made by a welder using a semiautomatic gun, the parameters and the procedure should have a greater tolerance.

Review Questions

Write your answers on a separate sheet of paper. Do not write in this book.
1. The GMAW process is considered either a semi-automatic or _____ automatic welding operation.
2. The hands-on skill of the welder is very important when welding in the _____ mode.
3. In GMAW, the two major parameters that must be established on the equipment before welding are _____ _____ and _____ _____.
4. Electrode extension is the length of wire that extends out of the contact tip. Increasing this length dimension _____ the amount of current flow, making the molten weld pool become sluggish.
5. Using the _____ technique for welding decreases penetration and increases spatter. It is always used for welding aluminum.
6. Using the _____ technique for welding increases penetration and has minimum spatter.
7. Gun angles used in GMAW are defined as "along the weld" or _____ angles.
8. Gun angles used in GMAW are defined as "across the weld" or _____ angles.
9. The weld bead pattern made when welding on the joint with little or no side-to-side motion, is commonly called a _____ _____.
10. The type of weld bead pattern made when welding on the joint with side-to-side motion is commonly called a _____ _____.
11. Information about the weld joint design and the welding parameters and variables may be placed on a form called a _____ _____.

The use of correct welding procedures and techniques will help assure the completed welds will meet all quality requirements of the fabrication specification. (American Welding Society)

CHAPTER 8 *Welding the Carbon Steels*

Objectives

After studying this chapter, you will be able to:
- Use the appropriate filler metal for different conditions.
- Prepare a joint for welding.
- Apply preheat, interpass temperature, and post-heat in correct proportions on various joints.
- Follow procedures for tooling operations.
- Employ welding procedures in sequence.
- Spot weld carbon steel.
- Plug and slot weld carbon steel.

Base Materials

Many types and grades of steel are included in the basic steel family. Carbon steels are magnetic and melt at approximately 2500°F. They are identified as a group of steels which contain:
- Carbon—1.70% maximum.
- Manganese—1.65% maximum.
- Silicon—0.60% maximum.
- Carbon steels may be further classified as:
- Low-carbon steel—up to 0.14% carbon.
- Mild carbon steel—0.15% to 0.29% carbon.
- Medium-carbon steel—0.30% to 0.59% carbon.

Some steels have been modified for use as machining stock by adding special materials to permit the machining operation. These steels may be called *free machining steels* or have other names denoting their use. Note: These materials are not weldable.

Low-alloy Steels

Low-alloy steels contain varying amounts of carbon in addition to many alloying elements. These elements include chromium, molybdenum, nickel, vanadium, and manganese. The elements increase the strength, toughness, and in some cases, the corrosion resistance.

Prior to the welding of these materials, either alone or in combination, the welding procedure should be qualified in conjunction with the thermal procedures used during welding and the heat treatment to be completed after welding.

Quenched and Tempered Steels

Quenched and tempered steels and high-strength steels (HSS) are produced by a number of steel companies. They have many trade names. These steels are hardened and tempered for specific mechanical values at the mill. By using these processes, the strength, impact resistance, corrosion resistance, and other properties can be improved. Manufacturers of these metals should be consulted prior to welding. This will assure proper welding so special attributes can be maintained.

Chrome-moly Steels

The alloys known as the *chromium-molybdenum steels* are used in applications requiring high strength. These steels may be annealed, hardened, or tempered. Special procedures are required when welding these materials to yield the desired mechanical properties. Always qualify the weld joint design, filler material type, and the welding procedure by adequate testing prior to welding this material.

Filler Materials

The filler materials used for welding mild steels are defined in the American Welding Society specification ANSI/AWS A5.18. A code is used to identify different filler materials. A typical example is:

E70S-1

The letter "E" stands for electrode. The number "70" means the wire has 70,000 pounds of tensile strength. The letter "S" identifies the wire as a solid wire. The number "1" identifies the class of wire. Each class is defined by the properties a wire will produce when welded with a specific single shielding gas, or a combination of gases. The chemical compositions of various steel welding wires are shown in the reference section and include:

- **E70S-2.** This wire is heavily deoxidized, meaning that it contains agents to fight oxidation. It is designed for producing sound welds in all grades of carbon steel: killed, semikilled, and rimmed. Due to added deoxidants (aluminum, zirconium, and titanium), it can weld carbon steel with a rusty surface. Argon-oxygen, argon-carbon dioxide, and carbon dioxide shielding gases can be used. In general, an extremely viscous (not fluid) weld pool is produced. This makes E70S-2 an ideal wire for short-arc welding out-of-position. A high oxygen or carbon dioxide content helps improve the wetting action of the pool.

- **E70S-3.** This wire is one of the most widely used wires with GMAW. E70S-3 wires can be used with either carbon dioxide, argon-oxygen, or argon-carbon dioxide in killed and semikilled steels. Rimmed steels should be welded with only argon-oxygen or argon-carbon dioxide. High welding currents used with carbon dioxide shielding gas may result in low strength. Either single-pass or multipass welds can be made with this electrode wire. The tensile strength for a single-pass weld in thin-gauge, low- and medium-carbon steels exceeds the base material while ductility is adequate. In a multipass weld, the tensile strength will range between 65,000 and 85,000 psi depending on the base metal dilution and type of shielding gas. The weld pool is more fluid than that of the E70S-2. The E70S-3 has better wetting action and flatter beads. This wire has its greatest application on automobiles, farm equipment, and home appliances.

- **E70S-4.** Wire electrode of this classification contains a higher level of silicon than the E70S-3. This improves the soundness of semikilled steels and increases weld metal strength. E70S-4 performs well with argon-oxygen, argon-carbon dioxide, and carbon dioxide shielding gases. It also can be used with either spray or short-arc techniques. Structural steels such as A7, A36, common ship steels, piping, pressure vessel steels, and A515 Grades 55 to 70 are usually welded with this wire. Under the same conditions, weld beads are generally flatter and wider than those made with the E70S-2 or the E70S-3.

- **E70S-5.** In addition to silicon and manganese, these wires contain aluminum (Al) as a deoxidizer. Because of the high aluminum content, they can be used for welding killed, semikilled, and rimmed steel with carbon dioxide shielding gas at high welding currents. Argon-oxygen and argon-carbon dioxide may also be used. Short-circuiting type transfer should be avoided because of extreme pool viscosity. Rusty surfaces can be welded using this wire with little sacrifice in weld quality. Welding is restricted to the flat position.

- **E70S-6.** Of all the electrodes, this one is high in manganese content and has the most silicon. These elements serve as deoxidizers. Like the E70S-5, this wire yields quality welds on most carbon steels. This assumes the use of carbon dioxide shielding gas and high welding currents. Also used are argon-oxygen mixtures containing 5% or more oxygen on high-speed welding. Since this wire contains no aluminum, the short-art technique is possible using carbon dioxide or argon-carbon dioxide shielding gases. Like the E70S-4, the weld pool is quite fluid.

- **E70S-1B.** This wire contains silicon and manganese as deoxidants plus molybdenum for increased strength. Welds can be made in all positions with argon-carbon dioxide and carbon dioxide shielding gases. Argon-oxygen is permitted for the flat position. Maximum mechanical properties are obtained with argon-oxygen and argon-carbon dioxide mixtures. Welding can be done over slightly rusted surfaces with some sacrifice in weld quality. This wire is mostly used for welding low-alloy steels such as AISI 4130.

Joint Preparation and Cleaning

Joint edges prepared by thermal cutting processes form a heavy oxide scale on the surface as in Figure 8-1. This oxide should be completely removed prior to welding. This prevents porosity and dross within the weld. To a great extent, GMAW nullifies foreign material on the base metal. However, good welds cannot be made over oxide scale.

Figure 8-1. The oxide scale and small gouge indentations were formed on this bevel cut by the oxyacetylene cutting process. The scale must be completely removed prior to welding.

Remove this scale and rough edges with a grinder. Foreign material next to the weld joint is removed by sanding, Figure 8-2. When making multiple passes, always clean the weld with a wire brush to remove oxides and foreign material. Where wire brushing does not remove the oxide scale, it must be removed by chipping or grinding. Failure to remove these oxides may result in lack of fusion when the next pass is made.

Many materials are received from the mill with a coating of oil to prevent rusting. When good-quality welds are desired, remove the oil with cleaning solvent before welding. This helps prevent porosity.

Preheating, Interpass Temperature, and Postheating

Preheating, interpass temperatures, and postheating temperatures are recommended for optimum mechanical properties, crack resistance, and hardness control on various types of carbon and low-alloy steels. *Preheating* is the application of heat to the weld area to obtain a specific temperature prior to the start of welding. The area should include the entire weld joint thickness and length.

Interpass temperature is the minimum/maximum temperature of the weld metal before the next pass in the multiple-pass weld is made. Interpass temperature should always be maintained until the postheating operation is started.

Postheating is the final temperature and time duration after welding is completed. This operation should be started immediately after the welding is done. Upon completion of postheating, the weld is allowed to cool to room temperature in still air.

In general, the carbon steels with less than 0.20% carbon and 1.0% manganese do not require any heating operation unless they are over approximately one-inch thick, they are to be welded in severe restraint, or the metal temperature is less than 50°F.

Figure 8-3. When using an oxyacetylene torch for heating, do not use an excess acetylene flame adjustment. Always use a neutral flame.

Carbon steels over 0.020% carbon and 1.0% manganese are more sensitive to cracking. The increase in carbon and manganese, degree of restraint, or an increase in weld thickness may require preheat, interpass temperature control, and finally postheat operations.

Carbon steels that contain carbon over 0.030% are often very crack sensitive and require heating and cooling temperatures and time rates. When establishing a weld procedure for these types of steels, the steel manufacturer and the wire manufacturer should be consulted for these temperatures.

The low-alloy steels such as the chrome-moly steels are often welded with the GMAW process. To prevent the formation of hard weld area grain structures, the amount of heating is dictated by the joint thickness, restraint of the weldment, and the final postheating. Welding procedures should be established using a 200°F to 400°F range to determine if cracking occurs. Cracks indicate the temperatures are too low and should be increased. Always allow any heated weldment to cool in still air or under a thermal blanket to slow down the cooling rate.

The *Quenched and Tempered steels* are a group of steels that have been heat treated for hardening and then tempered at a lower temperature to reduce the hardness and increase the ductility of the metal within specific tensile strength ranges. These steels are welded with the GMAW process. However, it is suggested that the manufacturer be contacted for the correct welding procedures to be used to maintain the desired mechanical values after welding.

Applying heat to the weld area is often accomplished with an oxyacetylene torch as shown in Figure 8-3. A temperature-measuring instrument may be used to indicate the temperature. Crayons or pellets shown in Figure 8-4 may also be used and are applied to the weld area. When the crayon or pellet temperature is reached, melting occurs thus indicating the actual temperature. *Note: Do not place crayons or pellets on the weld joint as they may contaminate the joint.*

Many weldments require another form of heating after the welding is completed to reduce the amount of

Figure 8-2. The material has been sanded to a "bright metal" finish to prepare the part for welding. This type of sander is often called a "PG" wheel.

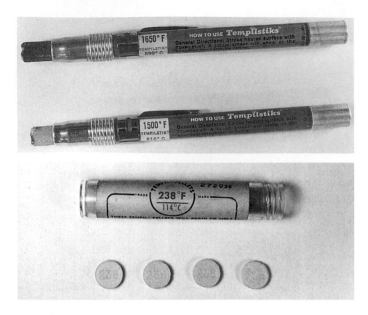

Figure 8-4. Temperature crayons and pellets may be used to determine preheating temperatures prior to welding.

Figure 8-5. Spray this material to prevent spatter build-up on the weldment. (York Mfg. Co.)

stress in the weld area. This is called a ***stress relief*** and is usually accomplished in a furnace where the entire part is placed for a period of time at a specific temperature. Due to the temperature tolerances and the extended time period for the operation, this cannot be done by a welder using a manual oxyacetylene torch.

Tooling

Tooling for GMAW is basically the same as for other welding processes. However, considerable spatter is generated during the GMAW short-arc mode, which can cause many problems within the tooling area. To overcome this problem, copper, aluminum, or stainless steel is generally used to minimize spatter pickup. These materials also reduce the possibility of ***magnetic arc blow*** (the altering of the arc pattern due to magnetism). In conjunction with the tooling materials, additional protection can be obtained from spatter build-up with the use of a liquid spray material which may be placed on the tooling areas where the spatter may fall. A typical spray material is shown in Figure 8-5.

The major items to be considered in designing tooling for the welding operation include:

- The proper weld joint alignment of the component parts.
- Heat control of the weld zone if required.
- Allowance for shrinkage of the weld joint during welding.
- Providing atmosphere to prevent contamination of the weld crown and the weld penetration if required.

- Assembly (loading) and disassembly (unloading) of the component parts.
- Positioning of the joint for welding.
- Accessibility for the welding gun and the welder to the weld joint.
- Tool inspection and maintenance program to maintain the integrity of the tool for the intended operation.

A typical tool used to hold materials for welding with the GMAW process is shown in Figure 6-38.

Welding Procedure

Establishing a welding procedure requires the consideration of many areas. The desired welding results can be obtained with careful consideration of these areas as applied to the specific operation.

1. Select the filler wire and the shielding gas or gas combination to match the properties of the material.
2. Select the process mode.
3. Select the correct process parameters for the material thickness from set-up charts in the reference section.
4. Establish the welding parameters on the power supply and the wire feeder.

5. Install filler material, check drive rolls for proper type and size, and install proper size contact tip and gas nozzle.
6. Select the proper shielding gas or gases and set the shielding gas flow rate (in cfh).
7. Set the shielding gas flow for the weld root side if required.
8. Establish the cleaning requirements.
9. Establish heating cycles, if required for the welding procedure.
10. Make a test weld on a piece of scrap material to determine that the weld is as desired. If tooling is designed for this operation, now is the time to test the tool and make any adjustments to be sure the tool is operating as designed. Adjust the welding equipment to fine-tune your welding parameters as you observe the welding operation.
    A. Establish tack weld procedures if required.
    B. Establish interpass temperatures to prevent cracking on multipass welds if required.
    C. Determine the proper interpass cleaning procedures to be used.
    D. Establish weld joint tolerances for fit up of the component parts.
    E. Use nondestructive test and destructive test to prove your final procedure and establish tolerances for inspection criteria.
11. During this period of weld testing, check your welding technique, focusing on the movement of the welding gun, and adjust your gun movement as required.
12. Determine if postheating is required. Establish temperatures and periods of time if needed.
13. Establish the tolerances for each parameter and variable of your procedure. Each part of the procedure will not be exact; therefore, a tolerance must exist for each of these areas.

Spot Welding the Carbon Steels

Spot welding may be used on a variety of metals as a primary method of joining components together. It may also be used to hold component parts assembled together for the final welding operation. The types and uses of this process are shown in Figure 8-6.

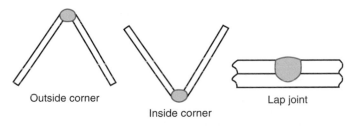

Figure 8-6. Uses of spot welding include corner and lap joints.

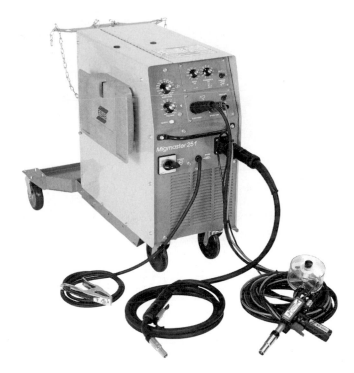

Figure 8-7. Welding machine controls include spot, stitch, and seam weld timers. (ESAB Welding & Cutting Products)

The spot welding process requires a welding machine with special timers installed to control the welding operation timing sequence. A typical welding power supply for thin-gauge materials with a timer assembly installed is shown in Figure 8-7. A typical timer assembly will include:

- **Welding timer.** This timer will control the length of time the welding current is on.
- **Shielding gas post-flow timer.** This timer controls the length of time the gas will flow after the arc is stopped. Usually the gas should flow for several seconds after the weld is stopped to prevent contamination of the molten metal.
- **Anti-stick timer** (commonly called a *burn-back timer*). The timer is placed into the welding circuit to allow the electrical current to flow to the welding wire after the welding wire stops feeding. The flow of current continues to "burn back" the welding wire to prevent the wire end from freezing into the molten pool.

The welding gun will require special nozzles with slots for the escape of welding heat, gases, and air during the welding operation. Each gun manufacturer will have special nozzles designed for these types of operations and they are not interchangeable with other types of manufacturers' guns. The various types of nozzles are shown in Figures 8-8, 8-9, and 8-10.

Figure 8-8. The flat edge of the nozzle is held tight against the top plate during the welding operation. The hot gases escape through the slots.

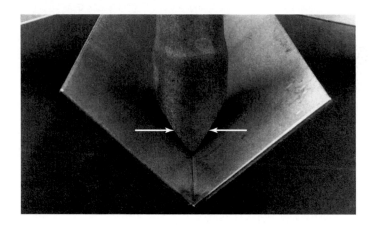

Figure 8-9. The inside corner nozzle is made to allow the gases to escape along these edges (indicated with the arrows) of the nozzle tip.

Figure 8-10. An outside corner nozzle has slots for the escape of the hot gases.

Pre-weld Cleaning and Fit-up

Lap joints often require special cleaning procedures if the weld is to be used in applications where strength is required. The area where the two metals contact must be clean. Since the heat for fusion must come through the upper material, any scale or rust will prevent fusion between the two plates. Materials should be cleaned with a sander as shown in Figure 8-2, sandblasted, or abrasively cleaned to bright metal. Any oil on the surface must be removed prior to welding.

Another problem that may be encountered is "gapping" of the two plates. If the plates are gapped, molten metal will cool before fusion is complete with the bottom plate and insufficient fusion will lower the strength of the weld.

When making spot welds on lap joints of different thicknesses, the thinnest material must always be on the top of the joint. If the design of the assembly cannot be changed, it is suggested that a plug weld be made in place of the spot weld.

Where spot welds are used on inner or outer corner joints, the fit-up gap of the mating edges must be minimized or burnthrough will occur. Any gap which will allow the welding wire to enter will burn off when the arc starts and the burned off wire will become a whisker on the inside of the weld.

Welding Procedures

1. Select the filler wire and the shielding gas or gas combination for the physical properties of the base material. For the mild steels, use the E70S-4 or the E70S-6 filler wires. They contain more deoxidants to ensure sound welds without porosity.
2. Select the process mode.
3. Select the correct process parameters for the thickness of the upper plate from the set-up charts in the reference section.
4. Establish welding parameters on the power supply and the wire feeder.

5. Set the shielding gas or gas combination cfh and post-flow time.

6. Install the correct diameter contact tip of the correct length for the type of welding nozzle. *Note: The contact tip must not extend to the end of the nozzle; clearance must be maintained for the weld crown.*

7. Install the special-design spot weld nozzle.

8. Set the burnback time so the electrode does not melt back into the contact tip at the end of the weld operation or stick into the molten pool.

9. Apply spatter-protection spray to the inside of the nozzle.

10. Using a test material of the same thickness as the required weld, place the gun nozzle firmly against the top plate and make a short weld. The time on for this weld test should be very short to check the equipment and to verify that all is operating satisfactorily. Increase the weld time in short increments until you obtain penetration into or through the bottom plate. On thin materials, a visual check can be done by observing the bottom of the plate. On heavier materials, the weld must be cut through the center and the weld nugget inspected for depth of penetration. A tensile test may also be done to determine the actual shear strength of the nugget.

11. For more penetration of the nugget, increase wire feed, decrease welding voltage, or increase weld time. If penetration is too deep, reverse this process.

12. Clean the gas nozzle often with a tool similar to the unit shown in Figure 8-11. Always apply spatter-protection spray to reduce spatter build-up. Allowing spatter to build up on the gas nozzles will reduce the effectiveness of the shielding gas.

13. After several test welds have been made, check the end of the welding wire to determine if the wire is oxidized. If the wire is oxidized, cut the wire from the end and increase the post-flow gas time.

Figures 8-12, 8-13, and 8-14 show completed lap, inside corner, and outside corner spot welds, respectively.

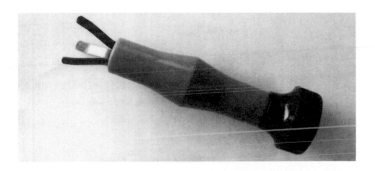

Figure 8-11. Cleaning tools are used often when making spot welds to remove spatter build-up within the nozzle.

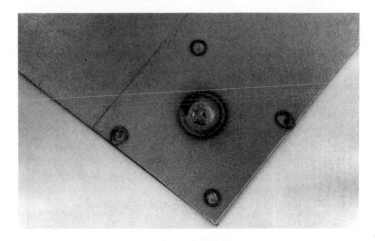

Figure 8-12. Completed lap joint spot weld.

Figure 8-13. Completed inside corner spot weld.

Figure 8-14. Completed outside corner spot weld.

Plug and Slot Welding the Carbon Steels

Plug and slot welds are often used in weldments where the material is too heavy for spot welds. They are also used where additional welding is required to reinforce the primary joint. Some typical uses of plug and slot welds are shown in Figure 8-15.

These welds are made with the same procedures used for seam welding except the welder must manipulate the gun around the hole or slot. Thin-gauge materials may be welded in one pass. Heavier materials, however, require several layers or passes. In these cases, always wire brush the completed weld before applying another pass or layer.

The major problem encountered in this type of weld is lack of fusion at the weld root. This is caused by an

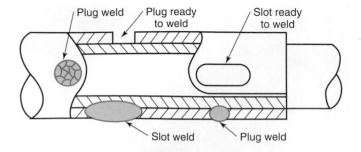

Figure 8-15. Plug and spot welds may be made in many combinations.

Figure 8-16. Completed plug weld.

improper gun angle as the welder tries to manipulate the gun around the hole. To minimize this condition, change the gun angles as the weld progresses. Practice on sample weldments to develop the proper techniques. A typical plug weld is shown in Figure 8-16.

Weld Defects

Do not feel frustrated when some of your welds contain defects. As your skills progress with practice, you will learn to avoid these mistakes. Chapter 11 discusses weld defects that are common when making various types of welds, regardless of the material. Refer to this chapter to learn the corrective action for the various defects you may encounter.

Chapter Review

Write your answers on a separate sheet of paper. Do not write in this book.

1. The carbon steels are _____ and melt at approximately 2500°F.
2. Carbon steels have a maximum of 1.70% of _____.
3. The quenched and tempered steels have been hardened and tempered to have specific _____ _____.
4. Most carbon steel electrodes are produced to a(n) _____ _____ _____ specification.
5. What does each digit in E70S-2 stand for?
6. What complications can be caused by welding over oxides?
7. When welding multipass welds, always remove surface _____ before depositing additional passes or layers.
8. Coatings of _____ on steel to prevent rusting should be removed with solvent prior to welding.
9. Why should low-alloy steels be preheated?
10. One of the major problems encountered in welding with steel tooling is the deflection of the arc by _____ _____ _____.
11. Lap joints must be cleaned before spot welding to _____ _____ prior to assembly and welding. When using various thicknesses of metal, the _____ is always on top.
12. Spot welds on lap joints with thin materials may be _____ examined to determine weld penetration.
13. You have welded a spot weld and the penetration is not sufficient. To apply more heat to the weld, you should _____ the wire feed speed, decrease the _____, or adjust the weld timer for a _____ period of time.
14. The major problem encountered in slot welding is _____ _____ _____ at the root of the weld.
15. The welder needs considerable _____ to avoid lack of fusion when slot welding.

CHAPTER 9 *Welding Stainless Steels*

Objectives

After studying this chapter, you will be able to:
- Select the proper filler wire for different types of stainless steel.
- Prepare a stainless steel joint for welding.
- Use appropriate tooling for stainless steels.
- Follow proper welding procedures for stainless steel.
- Spot weld stainless steel.
- Plug and slot weld stainless steel.

There are many industrial applications for welding stainless steels. In order to meet the vast demands, a variety of stainless steels have evolved. This chapter, however, will only discuss the austenitic group. When welding other stainless steels with GMAW, contact the manufacturer for recommendations or exact specifications and guidelines.

Base Materials

The stainless steel family is made up of several different groups of materials with many different types of materials within each family. The stainless steel groups are:
- Austenitic stainless steels.
- Ferritic stainless steels.
- Martensitic stainless steels.

Each material has specific welding procedures. However, the *austenitic stainless steel* is the most common in use today. *Austenite* is a metallurgical term identifying a specific type of grain structure. The materials within this austenitic stainless steel group are identified by the American Iron and Steel Institute as the 300 series (AISI 300). Figure 9-1 identifies the major austenitic materials.

The main elements of austenitic stainless steels are chromium and nickel. Chromium content in this type of stainless steel is 15-32% and nickel content is 8-37%. These elements help the stainless steel resist corrosion.

The austenitic group is used more than any of the other steel series. It has good corrosion resistance, excellent strength at both low and high temperatures, and has a high degree of toughness. Austenitic stainless steel cannot be hardened by heat treatment, and is also nonmagnetic.

Filler Materials

Various filler materials are used to join the alloys. The filler materials should be compatible with the base material. They should provide crack-resistant deposits, which are equal to or better than the base material in terms of soundness, corrosion resistance, strength, and toughness. Since weld metal with a composition equal to the base metal does not usually match the properties or performance of wrought base metal, it is common to enrich weld metal alloy content. The welding filler material usually used in the welding of these materials is shown in Figure 9-2. When selecting stainless steel filler material, always use a low-carbon (LC) or extra-low-carbon (ELC) filler wire when specified.

The filler material should be protected from contamination at all times. When the wire is installed on a machine, a spool cover should always completely enclose the material. Wire removed from the machine should be stored in a container to prevent the exposure to foreign material.

The common filler materials used for welding the austenitic stainless steels are defined by the American Welding Society specification ANSI/AWS A5.9, *Bare Stainless Steel Welding Electrodes and Rods*. The following types of filler wire can be used for stainless steel welding:
- **ER308.** This wire electrode is commonly used to weld type 304 stainless steel or any base material containing approximately 19% chromium and 9% nickel.

Commercially Wrought Stainless Steel Identification (AISI)

Type	C	Mn	Si	Cr	Ni	P	S	Others
				Composition, Percent[a]				
302	0.15	2.00	1.00	17.0-19.0	8.0-10.0	0.045	0.03	
302B	0.15	2.00	2.0-3.0	17.0-19.0	8.0-10.0	0.045	0.03	
303	0.15	2.00	1.00	17.0-19.0	8.0-10.0	0.20	0.15 min	0-0.6 Mo
303Se	0.15	2.00	1.00	17.0-19.0	8.0-10.0	0.20	0.06	0.15 Se min
304	0.08	2.00	1.00	18.0-20.0	8.0-10.5	0.045	0.03	
304L	0.03	2.00	1.00	18.0-20.0	8.0-12.0	0.045	0.03	
305	0.12	2.00	1.00	17.0-19.0	10.5-13.0	0.045	0.03	
308	0.08	2.00	1.00	19.0-21.0	10.0-12.0	0.045	0.03	
309	0.20	2.00	1.00	22.0-24.0	12.0-15.0	0.045	0.03	
309S	0.08	2.00	1.00	22.0-24.0	12.0-15.0	0.045	0.03	
310	0.25	2.00	1.50	24.09-26.0	19.0-22.0	0.045	0.03	
310S	0.08	2.00	1.50	24.0-26.0	19.0-22.0	0.045	0.03	
314	0.25	2.00	1.5-3.0	23.0-26.0	19.0-22.0	0.045	0.03	
316	0.08	2.00	1.00	16.0-18.0	10.0-14.0	0.045	0.03	2.0-3.0 Mo
316L	0.03	2.00	1.00	16.0-18.0	10.0-14.0	0.045	0.03	2.0-3.0 Mo
317	0.08	2.00	1.00	18.0-20.0	11.0-15.0	0.045	0.03	3.0-4.0 Mo
317L	0.03	2.00	1.00	18.0-20.0	11.0-15.0	0.045	0.03	3.0-4.0 Mo
321	0.08	2.00	1.00	17.0-19.0	9.0-12.0	0.045	0.03	5 x %C Ti min
329	0.10	2.00	1.00	25.0-30.0	3.0-6.0	0.045	0.03	1.0-2.0 Mo
330	0.08	2.00	0.75-1.5	17.0-20.0	34.0-37.0	0.04	0.03	
347	0.08	2.00	1.00	17.0-19.0	9.0-13.0	0.045	0.03	C
348	0.08	2.00	1.00	17.0-19.0	9.0-13.0	0.045	0.03	0.2 Cu[b,c]
384	0.08	2.00	1.00	15.0-17.0	17.0-19.0	0.045	0.03	

a. Single values are maximum unless indicated otherwise. b. (Cb + Ta) min — 10 x %C. c. Ta — 0.10% max.

Figure 9-1. American Iron and Steel Institute (AISI) identifies stainless steel by a numbering system and specifies the composition of each type. Note the following abbreviations for the elements listed in the chart: carbon (C), manganese (Mn), silicon (Si), chromium (Cr), nickel (Ni), phosphorous (P), sulfur (S), molybdenum (Mo), selenium (Se), titanium (Ti) and copper (Cu).

		Weld Filler Material									
		308	308L	309	310	316	316L	317	317L	330	347
Base Material		301 302		309	310						321 347
		304	304L 310S	309S		316	316L	317	317L	330	
		304N									348
				384		316N					
		308									

Figure 9-2. The filler wire listed at the top of each column is used to weld the base materials listed below it.

- **ER308L.** Wires of this type can be used for the same materials as ER308. The chromium and nickel contents are identical to ER308, but the lower carbon content reduces any possibility of carbide precipitation and the intergranular corrosion that can occur.
- **ER308LS.** This electrode has chemistry similar to the ER308 wire. However, a higher silicon level improves the wetting characteristics of the weld metal, particularly when argon-1% oxygen shielding gas is used. If the dilution of the base plate is extensive, high silicon content can cause greater crack sensitivity than a lower silicon content. This results from the weld being fully austenite or a low ferrite.

- **ER309.** This wire is used to weld type 309 and 309S stainless steel. It can be used to weld type 304 stainless steel where severe corrosion conditions will be encountered and for joining mild steel to type 304.
- **ER316.** This electrode is used to weld types 316 and 319 stainless steels. The addition of molybdenum makes this wire electrode useful for high-temperature service where creep resistance is desired.
- **ER316L.** This wire, due to the lower carbon content, will be less susceptible to intergranular corrosion caused by carbide precipitation when used in place of ER316.

- **ER347.** This wire is much less subject to intergranular corrosion from carbide precipitation, as tantalum and/or columbium are added to act as stabilizers. It is used for welding base materials with similar chemistry and where high-temperature strength is required.

Joint Preparation and Cleaning

Standard weld joint designs can be utilized for GMAW applications when welding stainless steels. In some cases where thin metals are involved, backing bars are used to prevent oxidation of the weld penetration. In some cases where the two materials are not compatible, one of the materials must be cladded for a compatible weld. A V-groove joint of two different materials is shown in Figure 9-3.

Dirty weld joints lead to porosity and carbon pickup. Therefore, remove all dirt, grease, crayon marks, etc., before assembly of the parts to be joined.

When wire brushing is required prior to or after welding, always use a stainless steel wire brush. A carbon steel brush could induce contamination, thus causing surface rusting.

Tooling

Nonmagnetic tooling must always be used when welding to prevent arc blow. If magnetic tooling is used, it will interfere with the flow of the metal from the electrode and cause defects in the completed weld. Since stainless steel retains heat, the tooling must remove the heat to reduce warpage, distortion, and prevent carbide precipitation of the weld and heat-affected zone. *Carbide precipitation* is the movement of chromium out of the grain into the grain boundary. The depletion of chromium decreases the amount of corrosion protection in the material.

Copper is nonmagnetic and is a very good backing material since it removes heat from the stainless steel very rapidly. The tooling may be machined to receive the penetration when required. Shielding gas may be used to prevent oxidation of the weld root side if

desired. *Note: Always protect the molten metal on the root side of the weld to prevent the absorption of oxygen and nitrogen during the solidification and cooling period.* A fixture with copper inserts and a gas manifold for butt welds is shown in Figure 9-4. A fixture designed for hold-down bars is shown in Figure 6-38.

Another method of protecting the root side of the weld joint is to use a flux designed for this particular application, Figure 9-5. The flux powder is mixed with acetone or alcohol and made into a paste. The paste is applied to the root side of the joint with a brush. The alcohol or acetone will evaporate, leaving the flux

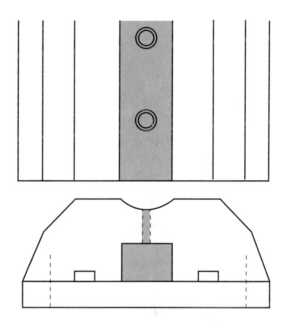

Figure 9-4. Backing bar design for admitting gas to the penetration side of the weld joint.

Figure 9-5. Typical commercial flux is often used where backing bars or tooling cannot be used.

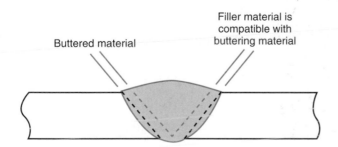

Figure 9-3. Where the materials to be joined are not compatible, one or both materials may be cladded (buttered) to make an acceptable weld with a common filler material.

adhering to the joint area. A weld joint with the paste applied is shown in Figure 9-6. The paste will melt and flow over the molten metal during the welding operation, thus preventing oxidation. After welding, the remaining flux may be removed by brushing the part in hot water or using a commercial acid cleaner.

Welding Procedures

Gas metal arc welding of stainless steel may be completed using any of the standard modes of metal transfer. These include:

- Short-arc transfer.
- Spray-arc transfer.
- Spray-arc pulse transfer.

Globular arc transfer is not generally used for stainless steel, since it is difficult to use in out-of-position welding.

The welding of stainless steels requires that the welding procedure be set up to avoid contamination of the molten weld pool. To accomplish this task, the gas nozzle must be of sufficient size to cover the weld, and the electrical stickout during welding must be held to procedure tolerances. The welding must not be done in drafty areas where gas coverage of the molten pool may be lost.

Figure 9-6. The flux powder remains on the part after the fluid evaporates.

Figure 9-7. A stainless steel tank assembly using a joggle weld joint design. Therefore, a backing gas or flux is not required to prevent oxidation of the weld root. (FluidTech Corp.)

Gas nozzles should be checked often to clear spatter inside of the nozzle which may interfere with the proper gas flow.

General welding parameters with the required shielding gas are listed in the Reference Section for both short-arc and spray-arc modes. The spray-arc pulsed mode is not referenced since different welding power supplies utilize different pulse frequencies. Refer to the welding machine manual for set-up parameters for this type of machine.

General Welding Conditions

As noted previously, welding procedures must be set up prior to welding to prevent contamination of the weld. Therefore, the following welding conditions should be checked prior to starting the weld.

- When installing a roll of wire on a machine that has been used for other materials, clean the entire wire feed system including the drive rollers and conduit liners.
- Always hold the gas nozzle as close to the work as possible.
- The forehand welding techniques form a flatter weld crown.
- Stringer beads reduce the overall heat input and the possibility of cracking.
- When making a multipass weld, always wire brush the previous pass with a stainless steel brush. *Note: Do not use stainless steel brushes which have been used on carbon steel material as they may pose contamination problems.*
- Do not leave craters at the end of the weld. Use tabs when automatic welding, and weld back over the end when using the semiautomatic mode. Craters are prone to cracking.

The stainless steel tank shown in Figure 9-7 is ready for automatic welding. A joggle-type joint design is used, which does not require a backing gas. The guns are offset to produce a flat weld crown on the completed weld as shown in Figure 9-8.

In Figure 9-9, stainless steel components and pipes have been welded to stainless steel tanks using the GMAW short-arc mode. The automatic spray-arc mode is shown in Figure 9-10.

Finished weldments may require cleaning to remove soot, oxides, and other materials. Commercial materials similar to the type shown in Figure 9-11 may be used for this purpose.

Spot Welding Stainless Steels

Select a filler wire type from the table shown in Figure 9-2 (or as specified) and set up the welding equipment. Remember to consider the type of process to

Figure 9-8. The completed weld is very smooth with good bead contour. The soot may be removed by wire brushing or with a chemical cleaner.

Figure 9-10. Automatic circular welds are very neat and the weld is consistent around the entire joint.

Figure 9-9. The weld technique for welding around the tubes must be developed to produce leak-free welds.

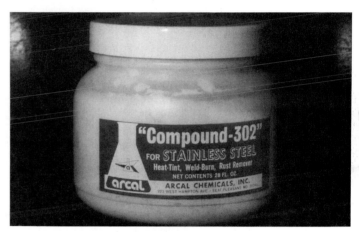

Figure 9-11. Commercial stainless steel weld cleaner. (Arcal Chemical, Inc.)

be used, the wire diameter, and the appropriate shielding gas.

The austenitic stainless steels are welded with the same equipment and procedures as the carbon steels (described in Chapter 8). Use a backing gas or a commercial flux to prevent oxidation of the weld penetration, whenever possible, when spot welding corner joints and lap joints where full penetration is required.

One of the major problems encountered in welding thin sections is distortion during the welding and cooling of each spot weld. Tooling may be required to contain warpage and maintain weld joint alignment.

Lap joints must be cleaned if an oxide scale is present. Wirebrush or use a commercial cleaner if fusion between the plates is to be complete.

Plug And Slot Welding Stainless Steels

Use the same parent metal and filler metal combinations as previously described, and set up your process with the proper gas for the welding mode. The welding techniques are the same as those used for welding the carbon steels.

The major problem encountered in this area is the condition of the holes or slots. Make sure that the hole or slot is free from burrs, slag (if plasma cut), or cutting oils which may have been used for drilling. Since these types of welds require good welding techniques to provide good fusion, always practice until your technique is correct.

Weld Defects

Chapter 11 describes many of the defects you may encounter when welding stainless steels with various types of weld joints. Examine the defect to determine why it happened and what can be done to prevent this from happening again. Then, change your procedure or technique to eliminate this condition.

Review Questions

Write your answers on a separate sheet of paper. Do not write in this book.

1. Chromium-nickel stainless steels are commonly called _____ stainless steels. This term also describes the grain structure of these materials.
2. Chromium-nickel stainless steels are identified by the _____ _____ _____ _____ as the 300 series.
3. Chromium-nickel stainless steels have very good _____ resistance.
4. The letters LC or ELC identify that the welding material must have a(n) _____-_____ or _____-_____-_____ content.
5. When brushing a stainless steel base material or weld, always use a(n) _____ _____ wire brush.
6. When welding stainless steel materials, _____ is not needed. If the material is overheated during the welding operation, _____ may result.
7. Nonmagnetic tooling reduces the possibility of _____ _____ _____ during welding.
8. When welding stainless steel along a seam, the _____ welding technique is used when making a fillet weld, to form a flatter weld crown.
9. The root side of full-penetration welds will be _____ if a gas or flux is not used during welding.
10. _____ _____ transfer is not generally used for welding stainless steel.
11. _____ _____ is the movement of chromium out of the grain into the grain boundary.
12. The use of _____ beads reduces the overall heat input and the possibility of cracking.
13. Craters occur at the ends of stainless steel welds. If they are not handled properly, _____ can occur.
14. A special _____ is used to remove heat lines along the weld.
15. Welding procedures for stainless steels are basically the same as for _____ steels.

CHAPTER 10

Welding Aluminum

Objectives

After studying this chapter, you will be able to:
- List eight characteristics of aluminum.
- Name the two basic groups of aluminum alloys and list many of them by number.
- Match the proper filler wire with various aluminum alloys for welding.
- Choose the correct filler wire to produce desired properties in a weld.
- Apply tooling to a joint for welding.
- Prepare an aluminum joint for welding.
- Demonstrate proper techniques for use in welding aluminum.

Aluminum is a *nonferrous metal* (one that contains no iron) that can be readily welded with the GMAW process. Some major characteristics of aluminum include:
- Thermal conductivity.
- Electrical conductivity.
- Ductility at subzero temperatures.
- Light weight.
- High corrosion resistance.
- Nonsparking.
- Nontoxic.
- No color change when heated.

Pure aluminum melts at approximately 1200°F (647°C); aluminum alloys from approximately 900°F to 1200°F (482°C to 647°C). An oxide film that forms on the surface of aluminum during manufacturing provides good corrosion resistance. But to produce quality welds, this film must be removed prior to welding.

Base Materials

The aluminum family is made up of pure aluminum and various alloy elements that provide the specific characteristics of the base material. With the addition of the alloying elements, each individual type is given a specific

designation for identification. Figure 10-1 shows the series designations for wrought aluminum alloy groups.

Nonheat-Treatable Wrought Alloys

The aluminum materials in this family include the 1XXX, 3XXX, 4XXX, and the 5XXX series. They contain specific alloy elements for strength; however, they do not respond to heat treatment for higher strength. Some of these materials will harden if cold-worked in the manufacturing state. Heat from the welding process may remove some of this strength in the heat-affected zone during welding.

Heat-Treatable Wrought Alloys

The heat-treatable alloys include materials in the 2XXX, 6XXX, and 7XXX series. Some 4XXX alloys are also heat-treatable. These materials are made in various heat-treated conditions and are identified by *temper designations,* as shown in Figure 10-2. Since these materials have higher strength in the annealed condition than the nonheat-treatable alloys, the weld zone may soften during welding. Heat-treatable alloys may require further heat treatment to regain their preweld condition.

Wrought Aluminum Alloy Groups	
Alloy Group	**Series Designation**
Aluminum, 99% minimum purity	1XXX
Aluminum-copper	2XXX
Aluminum-manganese	3XXX
Aluminum-silicon	4XXX
Aluminum-magnesium	5XXX
Aluminum-magnesium-silicon	6XXX
Aluminum-zinc	7000
Other	8000

Figure 10-1. Designations for wrought aluminum alloy groups.

Basic Temper Designations for Aluminum Alloys	
Designation	**Condition**
F	As fabricated.
O	Annealed.
H1	Strain-hardened only.
H2	Strain-hardened and partially annealed.
H3	Strain-hardened and thermally annealed.
W	Solution heat-treated.
T1	Cooled from elevated-temperature shaping process and naturally aged.
T2	Cooled from elevated-temperature shaping operation, cold-worked, and naturally aged.
T3	Solution heat-treated and then artificially aged.
T4	Solution heat-treated and naturally aged.
T5	Cooled from elevated-temperature shaping process, and then artificially aged.
T6	Solution heat-treated, then artificially aged.
T7	Solution heat-treated and stabilized.
T8	Solution heat-treated, cold-worked, then artificially aged.
T9	Solution heat-treated, artificially aged, and then cold-worked.
T10	Cooled from an elevated-temperature shaping process, cold-worked, and then artificially aged.

Figure 10-2. Temper designations for aluminum alloys.

Casting Alloys

Aluminum castings are often manufactured for pump housings, engine frames, motor bases, and similar uses. These alloys are identified by a different numerical system, as shown in Figure 10-3. Aluminum castings may be produced by sand mold, permanent mold, or die casting methods. These castings may either be nonheat-treatable, or heat-treatable with temper designations the same as those for wrought aluminum.

Filler Materials

Selecting the proper filler wire for welding aluminum will depend on many factors if the weld is to have quality, strength, and the ability to be used in the same applications as the base material. The common filler materials used for welding aluminum are defined by the specification AWS A5.10, *Aluminum and Aluminum Alloy Welding Rods and Bare Electrodes*.

Figure 10-4 lists the proper filler wire for welding various alloys, while filler wires for welding aluminum with specific properties are shown in Figure 10-5. Selection of the weld wire diameter will require consideration of such factors as the type of power supply, welding mode, type of wire feeder equipment, type of welding gun, joint design, position of the weld, and operator ability to manipulate the welding gun under adverse conditions if preheat is used.

Cast Aluminum Alloy Groups	
Alloy Group	**Series Designation**
Aluminum, 99% purity	1XX.X
Aluminum-copper	2XX.X
Aluminum-silicon-copper or aluminum-silicon-magnesium	3XX.X
Aluminum-silicon	4XX.X
Aluminum-magnesium	5XX.X
Aluminum-zinc	7XX.X
Aluminum-tin	8XX.X
Other alloy systems	9XX.X

Figure 10-3. Designations for cast aluminum alloy groups.

Filler materials should be stored and maintained under conditions that will prevent oxidation of the material. Oxide scale on the wire will cause defects in the weld, which could result in gas pockets and porosity.

Joint Design

Fillet welds do not generally present a problem when welding aluminum, since the weld shrinkage is not serious. In a *butt weld*, however, the center of the groove weld shrinks last, resulting in center-line cracking. This problem becomes more serious as the filler material combines with and dilutes into the base material at high temperatures. To reduce dilution of the filler wire and the base material, use a V-groove or U-groove weld whenever possible. Do not use the square-groove butt weld design–too much dilution will take place and the weld may be crack-sensitive.

Joint Preparation and Cleaning

When joint edges are prepared by the plasma arc cutting process, or cracks are removed by use of carbon arc gouging, a heavy *oxide scale* is formed on the base metal surface. All surfaces to be welded must be cleaned to bare metal prior to welding. Any oxidized metal remaining from cutting or gouging will remain in the weld, causing severe porosity. The heavy oxidation must be removed by machining or sanding. Several types of tungsten carbide and tool steel rotary files used for this purpose are shown in Figure 10-6. *Note: Do not use a grinding wheel to clean the oxide film from aluminum.*

Joint edges prepared by the *shearing* process should be sharp, without showing tearing or ridges. Where these conditions exist, dirt or oil may become entrapped, resulting in faulty welds. Straight edges of this type should be filed to clean metal.

Base Metal	6070	6061, 6063 6101, 6151 6201, 6951	5456	5454	5154 5254a	5086	5083	5052 5652a	5005 5050	3004 Alc. 3004	2219	2014 2024	1100 3003 Alc. 3003	1060 EC
1060, EC	ER4043h	ER4043h	ER5356c	ER4043e,h	ER4043e,h	ER5356c	ER5356c	ER4043i	ER1100c	ER4043	ER4145	ER4145	ER1100c	ER1100
1100, 3003 Alclad 3003	ER4043h	ER4043h	ER5356c	ER4043e,h	ER4043e,h	ER5356c	ER5356c	ER4043e,h	ER4043e	ER4043e	ER4145	ER4145	ER1100c	
2014, 2024	ER4145	ER4145									ER4145a	ER4145a		
2219	ER4043f,h	ER4043f,h	ER4043	ER4043h	ER4043h	ER4043i	ER4043	ER4043i	ER4043	ER4043	ER2319c,f,h			
3004 Alclad 3004	ER4043e	ER4043b	ER5356e	ER5654b	ER5654b	ER5356e	ER5356e	ER4043e,h	ER4043e	ER4043e				
5005, 5050	ER4043e	ER4043b	ER5356e	ER5654b	ER5654b	ER5356e	ER5356e	ER4043e,h	ER4043d					
5062, 5652a	ER5356b,c	ER5356b,c	ER5356b	ER5654b	ER5654b	ER5356e	ER5356e	ER5654a,b,c						
5083	ER5356e	ER5356e	ER5183e	ER5356e	ER5356e	ER5356e	ER5183							
5086	ER5386e	ER5356e	ER5356e	ER5356b	ER5356b	ER5356e								
5154, 5254a	ER5356b,c	ER5356b,c	ER5356b	ER5654b	ER5654a,b									
5454	ER5356b,c	ER5356b,c	ER5356b	ER5554a,b										
5456	ER5356e	ER5356e	ER5556e											
6061, 6063, 6101 6201, 6151, 6951	ER4043b,h	ER4043b,h												
6070	ER4043e,h													

Where no filler metal is listed, base metal combination is not recommended for welding.

a. Base metal alloys 5652 and 5254 are used for hydrogen peroxide service. ER5654 filler metal is used for welding both alloys for low-temperature service (150°F and below).
b. ER5183, ER5356, ER5554, ER5556, and ER5654 may be used.
c. ER4043 may be used.
d. Filler metal with same analysis as base metal is sometimes used.
e. ER5183, ER5356, or ER5556 may be used.
f. ER4145 may be used.
g. ER2319 may be used.
h. ER4047 may be used.
i. ER1100 may be used.

Figure 10-4. Wrought aluminum alloy filler material selection.

Base Metal	Filler Alloys(1) Preferred for Maximum As-welded Tensile Strength	Alternate Filler Alloys for Maximum Elongation	Base Metal	Filler Alloys(1) Preferred for Maximum As-welded Tensile Strength	Alternate Filler Alloys for Maximum Elongation
EC	1100	EC/1260	5086	5183	5183
1100	1100/4043	1100/4043	5154	5356	5183/5356
2014	4145	4043/2319 (3)	5357	5554	5356
2024	4145	4043/2319 (3)	5454	5554	5356
2219	2319	—	5456	5556	5183
3003	5183	1100/4043	6061	4043/5183	5356 (2)
3004	5554	5183/4043	6063	4043/5183	5183 (2)
5005	5183/4043	5183/4043	7039	5039	5183
5050	5356	5183/4043	7075	5183	—
5052	5356/5183	5183/4043	7079	5183	—
5083	5183	5183	7178	5183	—

The above table shows recommended choices of filler alloys for welds requiring maximum mechanical properties. For all special services of welded aluminum, inquiry should be made of your supplier.
1. Data shown are for "0" temper.
2. When making welded joints in 6061 or 6063 electrical conductor in which maximum conductivity is desired, use 4043 filler metal.

However, if strength and conductivity both are required, 5356 filler may be used and the weld reinforcement increased in size to compensate for the lower conductivity of the 5356 filler metal.
3. Low ductility of weldment is not appreciably affected by filler used. Plate weldments in these base metal alloys generally have lower elongations than those of other alloys listed in this table.

Figure 10-5. Filler materials used to obtain specific properties in completed welds.

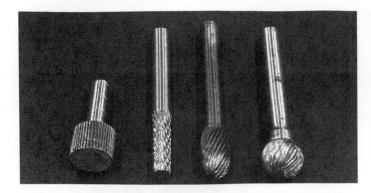

Figure 10-6. These types of files do not leave foreign metal in the aluminum base material.

Cleaning of joints in preparation for welding can be done by:

Chemical methods.
- Commercial degreasers.
- Commercial cleaning compounds. See Figure 10-7.
- Chemical baths.

Mechanical methods.
- Filing or scraping.
- Sanding with abrasive pads or abrasive wheels.
- Stainless steel brushes. *Note: Do not use carbon steel brushes or stainless steel brushes that have been used on carbon steel.*

After cleaning is completed, but before welding, wash the entire weld joint with alcohol or acetone. Allow to dry completely before welding. **Do not weld on a wet surface.**

This cleaning of the weld joint and the immediate weld area should be done immediately before welding. If welding must be delayed, the weld area should be covered with plain brown wrapping paper and sealed with tape to keep the joint clean.

Tooling

Tooling for backing is needed when making full penetration groove welds without an aluminum backing strip. Care should be taken to prevent the molten weld penetration from dropping through into air, since this will cause excessive oxidation and impede proper weld pool flow. Wherever possible, a solid aluminum or copper backing bar with a groove for the drop-through should be used. It will do an excellent job of preventing oxidation and establishing a base for the weld root.

Where high quality welds must be made, the backing tooling shown in Figure 10-8 may be used. This tool has a groove machined into its face for the weld drop-through and holes are drilled into a manifold in its base. The manifold is connected to a supply of argon or helium gas. The gas is admitted during welding to protect the weld drop-through from being contaminated. Hold-down bars are often used to secure the joint to the fixture; these may be either copper or stainless steel.

Figure 10-7. Commercial compounds actually etch the oxide film from the aluminum, leaving a very clean welding surface. (Arcal Chemicals, Inc.)

Welding Procedures

The short-arc, spray-arc, and spray-arc pulse modes may be used for the welding of aluminum. The short-arc mode is very difficult to use unless the procedure is very tightly controlled for the stickout requirement. This will generally require an automatic welding operation. With the advancement of shielding gas development, the spray-arc and spray-arc pulse modes offer lower overall current levels that can be used in all thicknesses and positions.

Argon is the main gas used for welding, since it provides good arc stability with the least amount of spatter. However, the addition of helium will often yield a hotter molten weld pool with better penetration and less porosity.

The shielding gas flow rates using a nozzle of 1/2″ to 5/8″ diameter should be approximately 25 to 40 cfh (cubic feet per hour). The higher gas flows should always be used with helium mixes. Always protect the arc from drafts and breezes.

Tables in the reference section of this book will be useful for the initial set-up of an aluminum welding procedure.

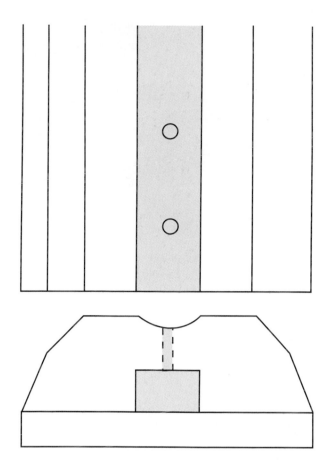

Figure 10-8. When argon gas is admitted to the lower section of the tool, it flows through the holes to the grooved area. The gas shields the penetration during welding and prevents the formation of oxides. This also assists in forming the root bead and prevents oxide folds.

Preheating is a common practice when welding aluminum. It reduces the amount of welding heat needed to achieve penetration, and increases welding speed. Oxyacetylene heating may be used, but the upper limit of heating should not exceed 300°F (149°C). This may be checked by using temperature-indicating crayons, pellets, or special indicating fluids placed on the part. These are shown in Figure 10-9. Be careful that the temperature-indicating material does not come into contact with the weld area.

When preheating, always remember that the welding procedure will require less welding current for the same amount of penetration, and the weld pool will be larger.

If you are making a *multipass weld*, always clean the completed weld with a stainless steel brush before applying the next pass. A typical stainless steel brush is shown in Figure 10-10.

When using the spray-arc mode on fillet or seam welds, the forehand welding technique will produce a good cleaning action of the weld joint. Figure 10-11 compares two welds—one made with the backhand technique, the other with forehand.

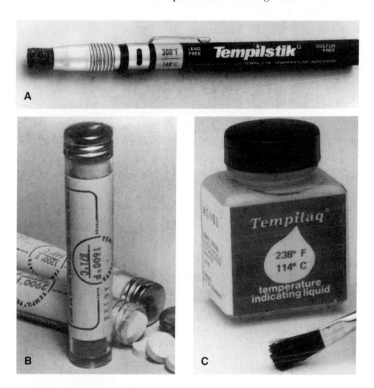

Figure 10-9. Temperature indicators. A—Crayon is moved across the heated metal, becoming fluid when indicated temperature is reached. B—Pellets are placed on the part being heated. They melt at the specified temperature. C—Temperature-indicating liquid is painted on the part being heated and melts at the indicated temperature.

You should avoid stopping the weld at the end of a joint and leaving a crater like the one shown in Figure 10-12. Aluminum is prone to cracking when the weld cross-section is thin. Using the same gun angle, always move back onto the end of the weld before stopping. This will fill the crater. Where this cannot be done, use a tab of parent metal tackwelded to the end of the joint, then run the weld over the end before stopping. Figure 10-13 shows tabs in place for the butt weld.

Spot Welding, Plug and Slot Welding

For these types of welds, use the procedures outlined in this chapter and specified in the reference section. Welding techniques for these joints are the same as those used with carbon steels and stainless steels.

Figure 10-10. A stainless steel wire brush like this one should be used to clean the joint after each pass when multipass welding on aluminum.

Figure 10-11. The dark-colored section of this weld was made with the backhand welding technique (note the soot on and around the bead). The light-colored section was made with the forehand welding technique. The clean weld bead and lighter metal color immediately next to the bead demonstrate the cleaning action of the reverse polarity welding current.

Figure 10-12. The crater at the left end of this weld was formed when welding stopped. The crater exhibits cracks and is very porous.

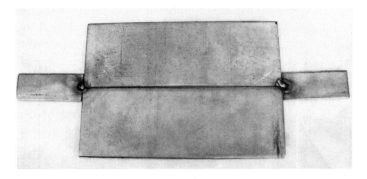

Figure 10-13. The runoff tabs used for this weld are approximately 3/4″ to 1″ wide and 2″ long.

Burnback

Burnback of the welding wire can be a very serious and costly problem when welding aluminum. Although burnback may occur in GMAW during any welding mode and with any type of wire, the problem is greatest when welding aluminum. This is due to the very low melting point (1200°F or 647°C) of the metal.

Figure 10-14 illustrates how the molten metal flows over the welding tip end and possibly inside the wire hole. This burnback is costly to repair; in many cases, the tip is ruined beyond repair. The following hints will help you avoid burnback when setting up or adjusting your procedure.

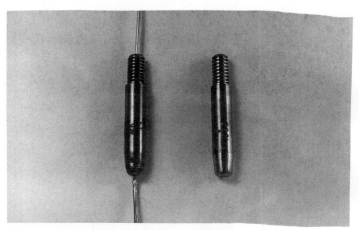

Figure 10-14. Aluminum filler wire has burned back over the end of the tip at left. This tip cannot be repaired to the original condition shown in the tip at right.

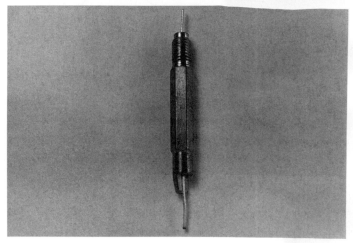

Figure 10-15. This welding tip is notched to permit access to the wire end. If burnback occurs, the wire can then be pried away from the tip and removed. Light filing of the tip end will remove any remaining filler metal.

- Start with a close arc voltage setting, then adjust to a higher desired setting.
- Start with a higher wire feed setting, then adjust to the lower desired setting.
- Use very short arc times when starting. Watch the arc very carefully until the correct values are established for the desired weld.
- Watch your wire supply very closely. If the supply runs out during a weld, burnback will occur and possibly ruin the contact tip.
- Keep your wire supply system in top shape. Clean out the liners on a regular basis and replace liners as necessary. Check drive rollers often and adjust as required.

The contact tip shown in Figure 10-15 is specially designed for use with aluminum. The cutaway section of the tip is designed to prevent burnback into the contact tip hole.

Weld defects

Do not feel frustrated when some of your welds contain defects. As your skills grow with practice, you will learn to avoid these mistakes. Chapter 11 discusses defects that commonly occur when making various types of welds, regardless of the material involved. Refer to that chapter to identify the corrective actions for the various types of defects you might encounter.

Review Questions

Write your answers on a separate sheet of paper. Do not write in this book.

1. List eight characteristics of aluminum.
2. What is a nonferrous material? Is the metal aluminum nonferrous?
3. The oxide film that forms on the surface of aluminum during manufacturing aids in preventing _____ of the base material during use.
4. Why must the oxide film be removed from the joint area before welding aluminum?
5. Common filler materials used for welding aluminum are defined by the specification _____ _____, *Aluminum and Aluminum Alloy Welding Rods and Bare Electrodes.*
6. The use of _____ filler material will have adverse effects on the quality of the completed weld.
7. Heavy oxide scale on the surface of a weld joint that has been thermally cut should be removed by _____ or sanding.
8. _____ _____ should never be used to remove scale from aluminum weld joints.
9. Commercial compounds or acids and _____ _____ wire brushes should be used to remove oxide films from joints or welds.
10. Final cleaning before welding may be done with _____ or _____ to remove contaminants.
11. What mode and technique are used most often when welding aluminum with GMAW?
12. The technique described in Question 11 has a(n) _____ _____ that removes surface oxides.
13. How does adding helium to the shielding gas affect weld quality?
14. How are craters avoided at the end of a weld pass?
15. Where possible, runoff _____ are used to reduce porosity at the end of the weld.

Test welds often must be made to establish a qualifed welding procedure before actual production welding begins. Qualifying the procedure will typically require subjecting test welds to both nondestructive and destructive testing procedures. Once a procedure is qualified, it can be used until a change in welding values requires requalification. (American Welding Society)

CHAPTER 11 Inspection, Defects, and Corrective Action

Objectives

After studying this chapter, you will be able to:
- Define a qualified welding procedure.
- Describe five nondestructive inspection methods.
- Troubleshoot defects for groove, fillet, plug, and spot welds.

Qualified Welding Procedure

Every gas metal arc weldment must meet a standard of quality for the particular weld used. Determining whether a standard of quality has been met can range from a casual look at a weld to see if the parts are joined, to a prescribed inspection for compliance with a specification. High-quality welds may require the use of test welds to establish correct parameters before production begins. Nondestructive testing is performed to verify quality, while destructive testing is done to verify mechanical values. If the tests are acceptable, the welding procedure is considered *qualified,* and the product can be welded using the procedure. If changes are made to the welding values after qualification, testing must be repeated to requalify the new procedure.

Remember: Any time you set up a machine, change the electrode, install a gas supply, or modify a parameter, run a test weld on a piece of scrap material. It is far better to throw away a piece of scrap metal than to ruin a production part.

Inspection Methods

All welds have flaws, or *discontinuities*. The role of inspection is to locate and determine the extent of those flaws. Extensive flaws are called *defects* and may cause a weld to fail. *Nondestructive testing* of a weld or assembly verifies quality and does not cause damage. The part is usable after the testing is done. *Destructive testing* is performed to determine the physical properties of a weld.

Only nondestructive inspection methods are discussed in this chapter. They include visual, liquid penetrant, magnetic particle, ultrasonic, and radiographic testing.

Visual Inspection

Visual inspection involves viewing the weld on the front or top surfaces (and penetration side, if visible) plus using rulers, scales, squares, and other tools to determine the condition of the weld. Common defects that can be identified by nondestructive visual examination include:
- Crown height.
- Crown profile.
- Underfill or low weld.
- Undercut.
- Overlap.
- Surface cracks.
- Crater cracks.
- Surface porosity.
- Weld size.
- Weld length.
- Joint mismatch.
- Warpage.
- Dimensional tolerances.
- Root side penetration.
- Root side profile.

The ripples in the weld should be even, without high and low areas or undercut at the weld toe. A special inspection tool can be used to check undercut depth, crown height on a butt weld, and drop-through on the penetration side of the butt weld. See Figure 11-1.

Fillet welds require the fillet leg be a specific size. Unless otherwise specified, the legs should be the same length, Figure 11-2. Fillet welds may also require inspection of the throat dimension, Figure 11-3.

Liquid Penetrant Inspection

Liquid penetrant inspection is a nondestructive test performed on the surface of a weld or the penetration side of a butt weld. Colored liquid dye or fluorescent

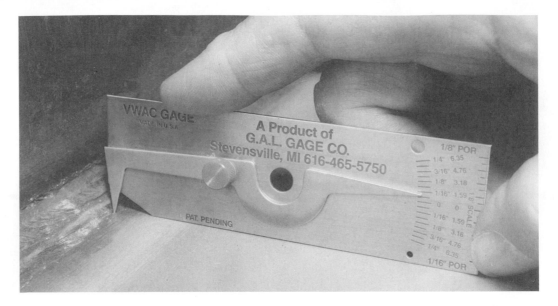

Figure 11-1. An inspection tool is used to check undercut at the edge of a weld. The amount of undercut is shown on the scale. (G.A.L. Gage Co.)

penetrant are applied to the weld. The penetrant seeps into any cracks or pits; excess penetrant is removed. Liquid developer is applied, drawing some of the penetrant out of the crevices. The dye permits defects to be seen. See Figure 11-4. Penetrant inspection does not reveal low welds or undercut. Another variation of this inspection technique uses a black light and fluorescent dye. Penetrant processes can be used in any position on metals, plastics, ceramics, or glass.

Magnetic Particle Inspection

Magnetic particle inspection is a nondestructive method of detecting the presence of cracks, seams, inclusions, segregations, porosity, lack of fusion, and similar discontinuities in magnetic materials. When a magnetic field is established in a ferromagnetic material (iron-based with magnetic properties) that contains one or more defects in the path of the magnetic flux, minute poles are set up at the defects. These poles have a stronger attraction for iron particles than does the surrounding material. The field is magnetized and a

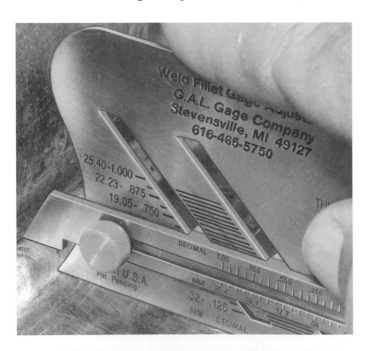

Figure 11-2. The pointer is placed at the edge of the weld leg. Leg size is indicated on the scale. (G.A.L. Gage Co.)

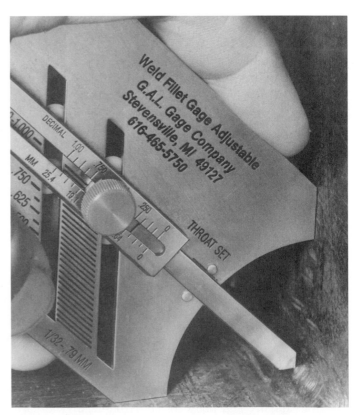

Figure 11-3. Throat size is used to determine concavity or convexity of the weld. (G.A.L. Gage Co.)

powder or fluid containing iron particles is applied. Any defects are shown by the pattern of the iron particles. See Figure 11-5. This process is mainly used for locating defects on the surface of the material; however, on thin materials the process is able to locate some defects below the surface.

Ultrasonic Inspection

Ultrasonic inspection is a nondestructive method of detecting the internal presence of cracks, inclusions, porosity, lack of fusion, and similar discontinuities in metals. High-frequency sound waves are transmitted through the part. The sound waves return to the sender and appear on a cathode ray tube (CRT). See Figure 11-6. Skilled technicians interpret the test results. Ultrasonic testing has certain advantages, including:

- Superb penetration power, permitting testing of thick materials and a variety of welds.
- Sensitivity sufficient to locate very small defects quickly.
- Ability to be done from one surface.

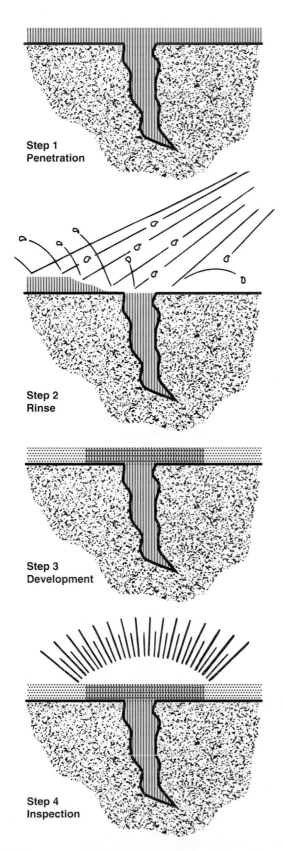

Figure 11-4. Liquid penetrant test sequence. (Magnaflux Corp.)

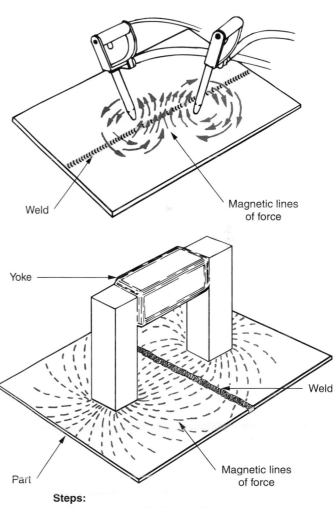

Steps:
1. Apply magnetic field using electric current.
2. Apply magnetic particles while power is on.
3. Blow away excess particles.
4. Inspect.

Figure 11-5. Magnetic particle test sequence.

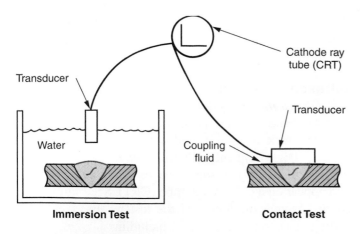

Figure 11-6. During an ultrasonic test, the part and a transducer are submerged in water. When this is not practical, the transducer is coupled (connected) to the test area by a thin layer of liquid.

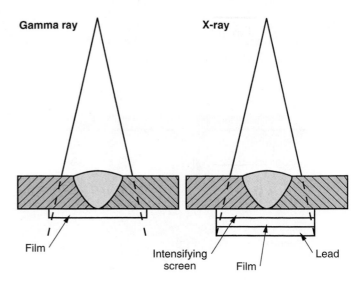

Figure 11-7. Radiographic test operation.

Radiographic Inspection

Radiographic inspection is another nondestructive test that shows the presence and nature of discontinuities in the interior of a weld. X-rays and gamma rays are used to penetrate the weld. Flaws are revealed on exposed radiographic film, Figure 11-7. Although radiographic inspection is expensive compared to other types of tests, the film creates a permanent record of the quality of a weld.

Weld Defects and Corrective Action

Certain defects are common to particular welds. Defects affecting groove, fillet, plug, and spot welds are described next, along with suggestions for correcting each of the problems. More than one corrective action may need to be taken for any given defect.

Groove Weld Defects

Lack of (incomplete) penetration. A weld that does not properly penetrate into the weld joint.

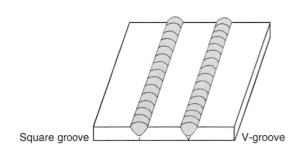

How to correct:
- Open groove angle.
- Decrease root face.
- Increase root opening.
- Increase amperage.
- Decrease voltage.
- Decrease travel speed.
- Change gun angle.
- Decrease stickout.
- Keep arc on leading edge of molten pool.

Lack of fusion. Fusion did not occur between the weld metal and fusion faces or adjoining weld beads.

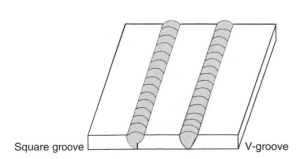

How to correct:
- Clean weld joint before welding.
- Remove oxides from previous welds.
- Open groove angle.
- Decrease root face.
- Increase root opening.
- Increase amperage.
- Decrease voltage.
- Decrease travel speed.
- Change gun angle.
- Decrease stickout.
- Keep arc on leading edge of molten pool.

Overlap. Weld metal that has flowed over the edge of the joint and improperly fused with the parent metal.

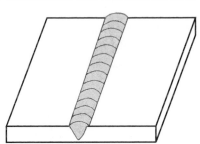

How to correct:
- Clean edge of weld joint.
- Remove oxides from previous welds.
- Reduce size of bead.
- Increase travel speed.

Undercut. Lack of filler material at the toe of the weld metal.

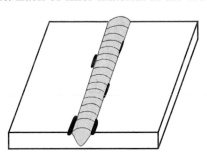

How to correct:
- Decrease travel speed.
- Increase dwell time at edge of joint on weave beads.
- Decrease voltage.
- Decrease amperage.
- Change gun angle.

Convex crown. A weld that is peaked in the center.

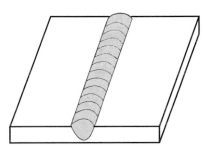

How to correct:
- Change gun angle.
- Increase current.
- Decrease stickout.
- Use weave bead technique with dwell time at edge of joint.

Craters. Formed at the end of a weld bead due to a lack of weld metal fill or weld shrinkage.

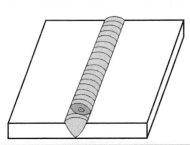

How to correct:
- Do not stop welding at end of joint (use tabs).
- Using same travel gun angles, move gun back on full section of weld before stopping.

Cracks. Caused by cooling stresses in the weld and/or parent metal.

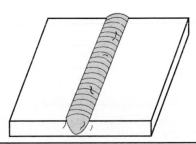

How to correct:
- Use wire with lower tensile strength or different chemistry.
- Increase joint preheat to slow the weld cooling rate.
- Allow joint to expand and contract during heating and cooling. Increase size of the weld.

Porosity. Caused by entrapped gas that did not have enough time to rise through the melt to the surface.

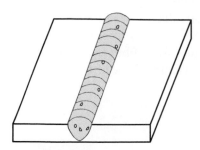

How to correct:

- Remove all heavy rust, paint, oil, or scale on joint before welding.
- Remove oxide film from previous passes or layers of weld.
- Check gas flow.
- Protect welding area from wind.
- Remove spatter from interior of gas nozzle.
- Check gas hoses for leaks.
- Check gas supply for contamination.

Linear porosity. Forms in a line along the root of the weld at the center of the joint where penetration is very shallow.

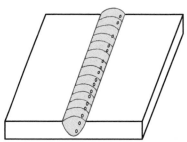

How to correct:

- Make sure root faces are clean.
- Increase current.
- Decrease voltage.
- Decrease stickout.
- Decrease travel speed.

Burn through. Occurs at a gap in the weld joint or place where metal is thin.

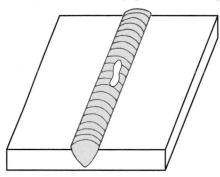

How to correct:

- Decrease current.
- Increase voltage.
- Increase stickout.
- Increase travel speed.
- Decrease root opening.

Whiskers. Pieces of weld wire that extend through the weld joint. Occurs where there is a root spacing that the wire slips into.

How to correct:

- Keep the arc on solidified metal and base metal.
- Do not allow the electrode to slip into the root opening.

Excessive penetration. Occurs when the weld metal penetrates beyond the bottom of the normal weld root contour.

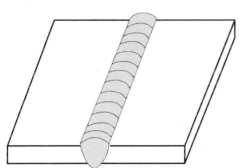

How to correct:
- Decrease root opening.
- Increase root face.
- Increase travel speed.
- Decrease amperage.
- Increase voltage.
- Change gun angle.
- Increase stickout.

Excessive spatter. Forms on the weld and parent metal.

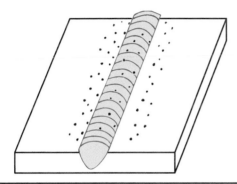

How to correct:
- Do not use CO_2 on steel.
- Change gun angle.
- Use pull technique
- Use antispatter spray.

Fillet Weld Defects

Lack of penetration. Insufficient weld metal penetration into the joint intersection.

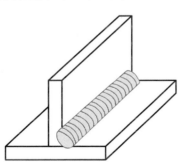

How to correct:
- Change gun angle.
- Increase amperage.
- Decrease voltage.
- Decrease size of weld deposit.
- Use stringer beads.
- Do not make weave beads on root passes.

Lack of fusion. Occurs in multiple-pass welds where layers do not fuse.

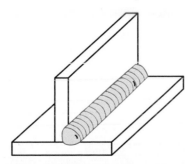

How to correct:
- Remove oxides and scale from previous weld passes.
- Increase amperage.
- Decrease voltage.
- Decrease travel speed.
- Change gun angle.
- Decrease stickout.
- Keep arc on leading edge of molten pool.

Overlap. Occurs in horizontal, multiple-pass fillet welds when too much weld is placed on the bottom layer.

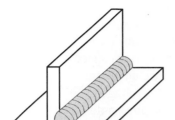

How to correct:
- Reduce the size of the weld pass.
- Reduce amperage.
- Change gun angle.
- Increase travel speed.

Undercut. Occurs at the top of the weld bead in horizontal fillet welds.

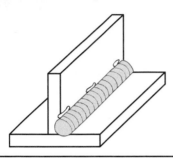

How to correct:
- Make a smaller weld.
- Make a multiple-pass weld.
- Change gun angle.
- Use smaller diameter electrode.
- Decrease amperage.
- Decrease voltage.

Convexity. A weld that has a high crown.

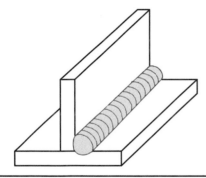

How to correct:
- Reduce amperage.
- Decrease stickout.
- Decrease voltage.
- Change gun angle.

Craters. Formed when weld metal shrinks below the full cross section of the weld.

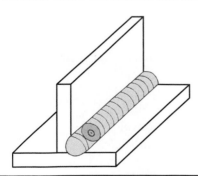

How to correct:
- Do not stop welding at the end of the joint (use tabs).
- Using the same travel gun angle, move the gun back on the full cross section before stopping.

Cracks. Occur in fillet welds just as they do in groove welds.

How to correct:
- Use suggestions for groove weld cracks.

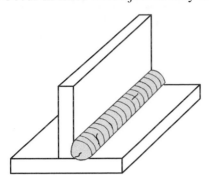

Burn through. Occurs when the molten pool melts through the base material and creates a hole.

How to correct:
- Decrease amperage.
- Increase travel speed.
- Change gun angle.

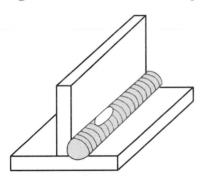

Porosity. Caused by entrapped gas that did not have enough time to rise through the melt to the surface.

How to correct:
- Use suggestions for groove weld porosity.

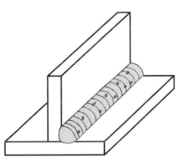

Linear porosity. Forms along the root of the joint interface.

How to correct:
- Use suggestions for groove weld linear porosity.

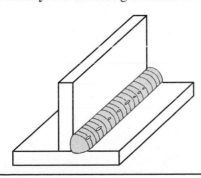

Plug Weld Defects

Lack of penetration. Occurs when the weld metal does not reach the proper depth in the bottom plate.

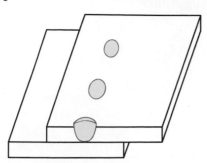

How to correct:
- Increase amperage.
- Decrease voltage.
- Decrease stickout.
- Start weld in center of plug hole, and fill it in a circular pattern.

Excessive penetration. Occurs when the root of the weld extends too far into the bottom sheet of the assembly.

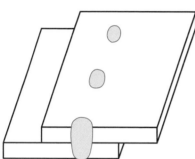

How to correct:
- Decrease amperage.
- Increase voltage.
- Increase stickout.
- Move gun in a circular pattern, and shorten weld operation.

Cracks. Occur at the center of the weld nugget and are caused by rapid cooling of weld metal.

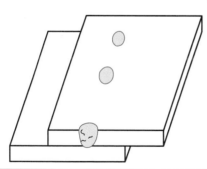

How to correct:
- Use an electrode of lower tensile strength or different chemistry.
- Increase preheat to slow cooling rate.
- Increase size of weld.

Porosity. Caused by entrapped gas that did not have enough time to rise through the melt to the surface.

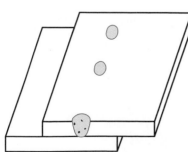

How to correct:
- Use suggestions for groove weld porosity.

Overlap. Occurs at the surface of the weld where too much metal has been deposited.

How to correct:
- Shorten welding time so less metal is deposited into the plug weld hole.

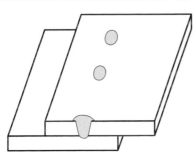

Craters. Occur in the center of the plug weld when not enough metal is placed in the hole during the weld cycle.

How to correct:
- Lengthen the weld cycle.

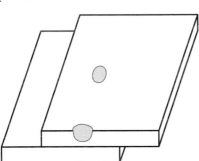

Spot Weld Defects

Lack of penetration. Occurs when the weld metal does not reach the proper depth in the bottom plate.

How to correct:
- Increase amperage.
- Increase weld time.
- Decrease voltage.
- Decrease stickout.

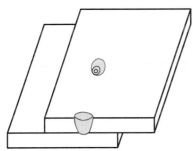

Excessive penetration. Occurs when molten metal penetrates through the bottom plate.

How to correct:
- Decrease amperage.
- Decrease weld time.
- Increase voltage.
- Increase stickout.

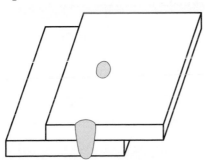

Porosity. Caused by entrapped gas that did not have enough time to rise through the melt to the surface.

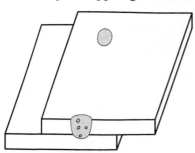

How to correct:
- Increase arc time.
- Increase amperage.
- Decrease voltage.
- Decrease stickout.
- Make sure the metal is clean before welding the assembly.

Cracks. Occur for the same reasons they do in plug welds.

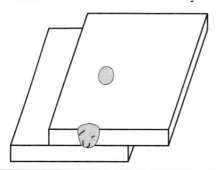

How to correct:
- Use suggestions for plug weld cracks.

Review Questions

Please do not write in this text. Write your answers on a separate sheet of paper.

1. What is a qualified welding procedure?
2. _____ inspection is done by looking at the weld to determine surface irregularities.
3. The tool shown below is used to measure the depth of _____ and to check _____ on butt welds.

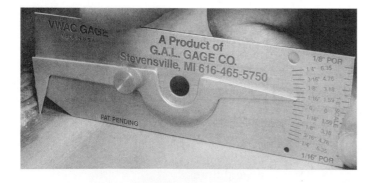

4. Fillet weld inspection includes the size of the _____ and the _____ dimension.

5. _____ inspection uses an electrical current and ferromagnetic material to detect discontinuities.
6. _____ inspection uses high-frequency sound waves for locating discontinuities.
7. Explain how radiographic inspection shows the presence of discontinuities in the interior of a weld.
8. When a weld does not penetrate completely through the base material, the condition is described as _____.
9. A weld that does not have sufficient cross-section at the end of the pass has a _____.
10. _____ is the formation of gas pockets in the weld body or on the crown surface.
11. Molten metal that is expelled from the weld pool is called _____.
12. _____ are caused by the shrinking of the weld and the weld area.
13. _____ affects fillet welds on thin metals and can be caused by welding with excessive amperage.
14. A weld that is peaked in the middle is called a _____.
15. A weld that is caved inward is called a _____.

CHAPTER 12

General Welding Procedures

Objectives

After studying this chapter, you will be able to:

- Make test welds with various designs, metals, and thicknesses.
- Produce welds using a GMAW schedule.

This chapter details the welding of differing types of steel, stainless steel, and aluminum, Figure 12-1. The included weld schedules have been made using existing shop procedures. They may vary considerably from the set-up charts shown in the reference section of the chapters regarding certain materials. Many factors exist in this welding process. Therefore, it is important to confirm your procedure by making *test welds*.

A test weld should have the following in common with the required weld: joint design, metal, and material thickness. In other words, copy the weld you are going to make. Upon completion, the test weld should be inspected for quality. The weld must perform its intended function. Testing can be done by visual inspection, cutting welds apart for a macro test, pulling until destruction, bending the welded joint in a press, applying a penetrant, or performing one of many other testing processes.

Once you have started welding, check your welds as they are made. The welds should look like your test welds. If they do not, stop and find out why. Minor changes in your schedule may have radical effects on completed welds. These changes are important to the quality of each weld. Tolerances should be established for each parameter or variable. Good welds can be produced by skilled welders using proven schedules within established tolerances. Faulty welds can be produced by the same welders using unproven schedules or tolerances.

Proven schedules and the resulting welds are shown in Figures 12-2 through 12-15.

Figure 12-1. Jigs are often used to hold parts together during welding. These aluminum frames are being welded with the short-arc mode using argon as the shielding gas. (ODL, Inc.)

GAS METAL ARC WELDING SCHEDULE

PROCESS MODE *SHORT-ARC*

WELD TYPE *FILLET*   POSITION *FLAT/HORIZONTAL*

BASE METAL TYPE *STEEL*   BASE METAL THICKNESS *1/16" - 1/16"*

WIRE TYPE *E70S-3*   WIRE DIA. *.035"*   WIRE SPEED (IPM) *170-180*

GAS TYPE *Ar-75%*   GAS TYPE *CO_2 - 25%*   FLOW RATE (CFH) *35*

NOZZLE TYPE *STD*   NOZZLE DIA. *3/8" - .375"*

CONTACT TIP TYPE (long) *X*   (short) ___   (standard) ___   DIA. *.035"*

AMPERES *100 - 110*   VOLTS (arc) *16 - 17*   STICKOUT *3/8"*

SPOTWELD TIME (seconds) ___   BURNBACK TIME (seconds) ___

TRAVEL (forehand) ___   (backhand) *X*   (uphill) ___   (downhill) ___

PRE-WELD CLEANING TYPE *WIRE BRUSH*

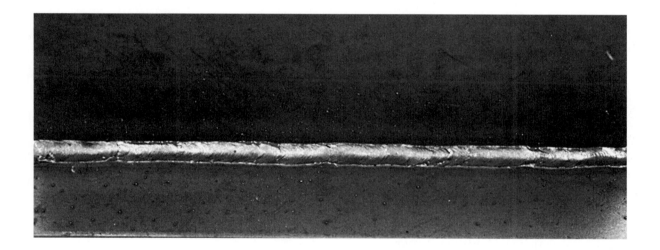

Figure 12-2.

GAS METAL ARC WELDING SCHEDULE

PROCESS MODE __SHORT-ARC__

WELD TYPE __EDGE__   POSITION __ALL__

BASE METAL TYPE __STEEL__   BASE METAL THICKNESS __1/16"__

WIRE TYPE __E70S-3__   WIRE DIA. __.030"__   WIRE SPEED (IPM) __210 - 220__

GAS TYPE __CO_2 - 100%__   GAS TYPE _____   FLOW RATE (CFH) __35__

NOZZLE TYPE __STD__   NOZZLE DIA. __3/8" - .375"__

CONTACT TIP TYPE (long) __X__ (short) ___ (standard) ___   DIA. __.030"__

AMPERES __85__   VOLTS (arc) __18 - 20__   STICKOUT __3/8"__

SPOTWELD TIME (seconds) ___   BURNBACK TIME (seconds) ___

TRAVEL (forehand) __X__ (backhand) ___ (uphill) ___ (downhill) __X__

PRE-WELD CLEANING TYPE __WIRE BRUSH__

Figure 12-3.

GAS METAL ARC WELDING SCHEDULE

PROCESS MODE *SHORT-ARC*

WELD TYPE *FILLET*   POSITION *HORIZONTAL*

BASE METAL TYPE *STEEL*   BASE METAL THICKNESS *1/8" - 1/8"*

WIRE TYPE *E70S-2*   WIRE DIA. *.035"*   WIRE SPEED (IPM) *270 - 280*

GAS TYPE *Ar-75%*   GAS TYPE *CO_2 - 25%*   FLOW RATE (CFH) *30*

NOZZLE TYPE *STD*   NOZZLE DIA. *3/8" - .375"*

CONTACT TIP TYPE (long) *X*   (short) ___   (standard) ___   DIA. *.035"*

AMPERES *140 - 150*   VOLTS (arc) *18 - 20*   STICKOUT *3/8"*

SPOTWELD TIME (seconds) ___   BURNBACK TIME (seconds) ___

TRAVEL (forehand) *X*   (backhand) ___   (uphill) ___   (downhill) ___

PRE-WELD CLEANING TYPE *WIRE BRUSH*

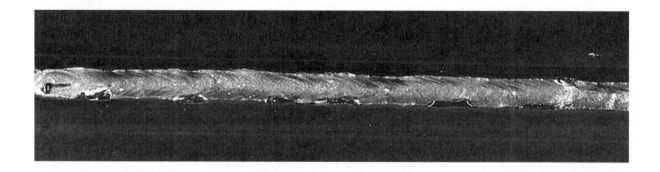

Figure 12-4.

GAS METAL ARC WELDING SCHEDULE

PROCESS MODE *SHORT-ARC*

WELD TYPE *FILLET/LAP*   POSITION   *FLAT/HORIZONTAL*

BASE METAL TYPE *STEEL*   BASE METAL THICKNESS *1/8"*

WIRE TYPE *E70S-4*   WIRE DIA. *.035"*   WIRE SPEED (IPM) *275 - 285*

GAS TYPE _____   GAS TYPE *CO_2 - 100%*   FLOW RATE (CFH) *35*

NOZZLE TYPE *STD*   NOZZLE DIA. *3/8" - .375"*

CONTACT TIP TYPE (long) *X*   (short) ___   (standard) ___   DIA. *.035"*

AMPERES *140 - 150*   VOLTS (arc) *20 - 22* STICKOUT *3/8"*

SPOTWELD TIME (seconds) ___   BURNBACK TIME (seconds) ___

TRAVEL (forehand) ___   (backhand) *X*   (uphill) ___   (downhill) ___

PRE-WELD CLEANING TYPE *WIRE BRUSH*

Figure 12-5.

GAS METAL ARC WELDING SCHEDULE

PROCESS MODE *SHORT-ARC*

WELD TYPE *FILLET*   POSITION *HORIZONTAL*

BASE METAL TYPE *STEEL*   BASE METAL THICKNESS *3/16" - 3/16"*

WIRE TYPE *E70S-2*   WIRE DIA. *.045"*   WIRE SPEED (IPM) *210 - 225*

GAS TYPE *Ar-75%*   GAS TYPE *CO_2 - 25%*   FLOW RATE (CFH) *35*

NOZZLE TYPE *STD*   NOZZLE DIA. *5/8" - .620"*

CONTACT TIP TYPE (long) *X* (short) ___ (standard) ___   DIA. *.045"*

AMPERES *210 - 215*   VOLTS (arc) *19 - 20*   STICKOUT *3/8"*

SPOTWELD TIME (seconds) ___   BURNBACK TIME (seconds) ___

TRAVEL (forehand) *X*   (backhand) ___   (uphill) ___   (downhill) ___

PRE-WELD CLEANING TYPE *WIRE BRUSH*

Figure 12-6.

GAS METAL ARC WELDING SCHEDULE

PROCESS MODE *SHORT-ARC*

WELD TYPE *GROOVE/OPEN*   POSITION *FLAT*

BASE METAL TYPE *STEEL*   BASE METAL THICKNESS *3/16"*

WIRE TYPE *E70S-2*   WIRE DIA. *.045"*   WIRE SPEED (IPM) *180 - 190*

GAS TYPE *Ar-75%*   GAS TYPE *CO_2 - 25%*   FLOW RATE (CFH) *30*

NOZZLE TYPE *STD*   NOZZLE DIA. *3/8" - .375"*

CONTACT TIP TYPE (long) *X*   (short) ___   (standard) ___   DIA. *.045"*

AMPERES *175 - 185*   VOLTS (arc) *18 - 20*   STICKOUT *3/8"*

SPOTWELD TIME (seconds) ___   BURNBACK TIME (seconds) ___

TRAVEL (forehand) ___   (backhand) ___   (uphill) ___   (downhill) ___

PRE-WELD CLEANING TYPE *WIRE BRUSH*

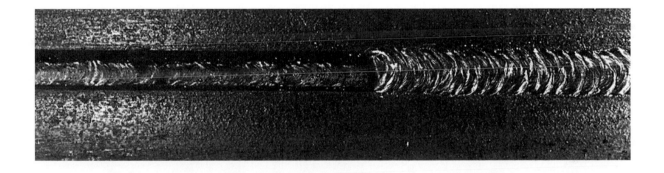

Figure 12-7.

GAS METAL ARC WELDING SCHEDULE

PROCESS MODE *SPRAY-ARC*

WELD TYPE *FILLET*   POSITION *FLAT*

BASE METAL TYPE *STEEL*   BASE METAL THICKNESS *3/16" - 3/16"*

WIRE TYPE *E70S-2*   WIRE DIA. *.035"*   WIRE SPEED (IPM) *320 - 340*

GAS TYPE *Ar - 98%*   GAS TYPE *O_2 - 2%*   FLOW RATE (CFH) *40*

NOZZLE TYPE *AIR COOLED*   NOZZLE DIA. *5/8" - .620"*

CONTACT TIP TYPE (long) ___ (short) *X* (standard) ___ DIA. *.035"*

AMPERES *180 - 190*   VOLTS (arc) *24 - 25*   STICKOUT *3/4" - 1"*

SPOTWELD TIME (seconds) ___   BURNBACK TIME (seconds) ___

TRAVEL (forehand) *X* (backhand) ___ (uphill) ___ (downhill) ___

PRE-WELD CLEANING TYPE *WIRE BRUSH*

Figure 12-8.

GAS METAL ARC WELDING SCHEDULE

PROCESS MODE *SPRAY-ARC*

WELD TYPE *GROOVE*   POSITION *FLAT*

BASE METAL TYPE *STEEL*   BASE METAL THICKNESS *1/2" - 1/2"*

WIRE TYPE *E70S-2*   WIRE DIA. *1/16"*   WIRE SPEED (IPM) *195 - 205*

GAS TYPE *Ar-98%*   GAS TYPE O_2 - *2%*   FLOW RATE (CFH) *40*

NOZZLE TYPE *AIR COOLED*   NOZZLE DIA. *5/8" - .620"*

CONTACT TIP TYPE (long) ___ (short) *X* (standard) ___ DIA. *1/16"*

AMPERES *320 - 330*   VOLTS (arc) *26 - 27*   STICKOUT *3/4" - 1"*

SPOTWELD TIME (seconds) ___   BURNBACK TIME (seconds) ___

TRAVEL (forehand) ___   (backhand) *X* (uphill) ___ (downhill) ___

PRE-WELD CLEANING TYPE *WIRE BRUSH*

Figure 12-9.

GAS METAL ARC WELDING SCHEDULE

PROCESS MODE *SPRAY-ARC*

WELD TYPE *PLUG*   POSITION *FLAT*

BASE METAL TYPE *STEEL*   BASE METAL THICKNESS *3/16"*

WIRE TYPE *E70S-3*   WIRE DIA. *.035"*   WIRE SPEED (IPM) *370*

GAS TYPE *Ar - 98%*   GAS TYPE *O_2 - 2%*   FLOW RATE (CFH) *40*

NOZZLE TYPE *AIR COOLED*   NOZZLE DIA. *5/8" - .620"*

CONTACT TIP TYPE (long) ___   (short) *X*   (standard) ___   DIA. *.035"*

AMPERES *190*   VOLTS (arc) *25 - 26* STICKOUT *3/4" - 1"*

SPOTWELD TIME (seconds) ___   BURNBACK TIME (seconds) ___

TRAVEL (forehand) ___   (backhand) ___   (uphill) ___   (downhill) ___

PRE-WELD CLEANING TYPE *WIRE BRUSH*

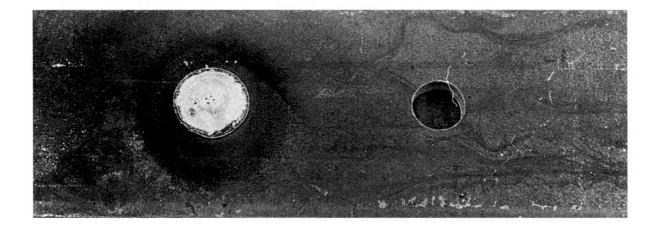

Figure 12-10.

GAS METAL ARC WELDING SCHEDULE

PROCESS MODE *SHORT-ARC*

WELD TYPE *FILLET*   POSITION *HORIZONTAL*

BASE METAL TYPE *ALUMINUM*   BASE METAL THICKNESS *.080" - .080"*

WIRE TYPE *ER4043*   WIRE DIA. *.035"*   WIRE SPEED (IPM) *175 - 185*

GAS TYPE *ARGON*   GAS TYPE _____ % FLOW RATE (CFH) *30*

NOZZLE TYPE *STD*   NOZZLE DIA. *3/8" - .375"*

CONTACT TIP TYPE (long) *X* (short) ___ (standard) ___   DIA. *.035"*

AMPERES _____   VOLTS (arc) *17 - 20* STICKOUT *3/8"*

SPOTWELD TIME (seconds) ___   BURNBACK TIME (seconds) ___

TRAVEL (forehand) *X* (backhand) ___ (uphill) ___ (downhill) ___

PRE-WELD CLEANING TYPE *NONE*

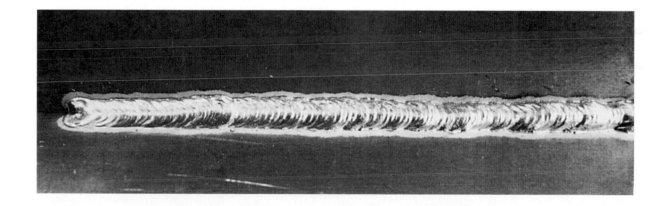

Figure 12-11.

GAS METAL ARC WELDING SCHEDULE

PROCESS MODE *SPRAY-ARC*

WELD TYPE *FILLET*   POSITION *FLAT*

BASE METAL TYPE *ALUMINUM*   BASE METAL THICKNESS *1/8" - 1/8"*

WIRE TYPE *ER4043*   WIRE DIA. *.035"*   WIRE SPEED (IPM) *350*

GAS TYPE *ARGON*   GAS TYPE _____ %   FLOW RATE (CFH) *35*

NOZZLE TYPE *STD*   NOZZLE DIA. *1/2" - .500"*

CONTACT TIP TYPE (long) ___ (short) *X* (standard) ___ DIA. *.035"*

AMPERES *110*   VOLTS (arc) *21*   STICKOUT *1/2" - 3/4"*

SPOTWELD TIME (seconds) ___   BURNBACK TIME (seconds) ___

TRAVEL (forehand) *X* (backhand) ___ (uphill) ___ (downhill) ___

PRE-WELD CLEANING TYPE *S/STEEL WIRE BRUSH*

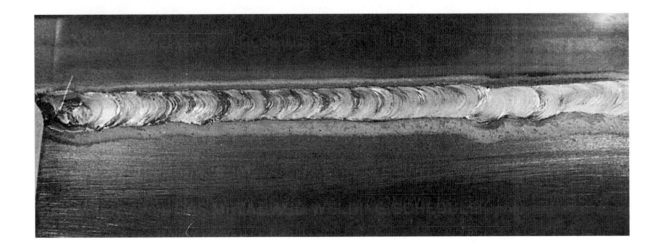

Figure 12-12.

GAS METAL ARC WELDING SCHEDULE

PROCESS MODE *SPRAY-ARC*

WELD TYPE *FILLET*   POSITION *HORIZONTAL*

BASE METAL TYPE *ALUMINUM*   BASE METAL THICKNESS *1/4"*

WIRE TYPE *ER5356*   WIRE DIA. *3/64"*   WIRE SPEED (IPM) *350*

GAS TYPE *ARGON*   GAS TYPE _____ %   FLOW RATE (CFH) *40*

NOZZLE TYPE *STD*   NOZZLE DIA. *5/8" - .620"*

CONTACT TIP TYPE (long) ___   (short) *X*   (standard) ___   DIA. *3/64"*

AMPERES *190*   VOLTS (arc) *24*   STICKOUT *3/4" - 1"*

SPOTWELD TIME (seconds) ___   BURNBACK TIME (seconds) ___

TRAVEL (forehand) ___   (backhand) ___   (uphill) ___   (downhill) ___

PRE-WELD CLEANING TYPE *DEGREASE – WIRE BRUSH*

Figure 12-13.

GAS METAL ARC WELDING SCHEDULE

PROCESS MODE _SHORT-ARC_

WELD TYPE _JOGGLE_   POSITION _ROTATE-AUTOMATIC_

BASE METAL TYPE _S/STEEL_   BASE METAL THICKNESS _14 GA._

WIRE TYPE _ER316L_   WIRE DIA. _.035"_   WIRE SPEED (IPM) _240 - 260_

GAS TYPE _TRI-MIX_   GAS TYPE _____ %   FLOW RATE (CFH) _40_

NOZZLE TYPE _STD_   NOZZLE DIA. _5/8" - .620"_

CONTACT TIP TYPE (long) _X_ (short) ___ (standard) ___   DIA. _.035"_

AMPERES _90 - 130_   VOLTS (arc) _20 - 21_   STICKOUT _3/8"_

SPOTWELD TIME (seconds) ___   BURNBACK TIME (seconds) ___

TRAVEL (forehand) _X_ (backhand) ___ (uphill) ___ (downhill) ___

PRE-WELD CLEANING TYPE ___WIRE BRUSH – DEGREASE___

Figure 12-14.

GAS METAL ARC WELDING SCHEDULE

PROCESS MODE _SPRAY-ARC_

WELD TYPE _FILLET_   POSITION _FLAT_

BASE METAL TYPE _S/STEEL_   BASE METAL THICKNESS _3/16" - 3/16"_

WIRE TYPE _ER308L_   WIRE DIA. _.035"_   WIRE SPEED (IPM) _410_

GAS TYPE _Ar - 98%_   GAS TYPE _O_2 - 2%_   FLOW RATE (CFH) _40_

NOZZLE TYPE _AIR COOLED_   NOZZLE DIA. _5/8" - .620"_

CONTACT TIP TYPE (long) ___   (short) _X_ (standard) ___   DIA. _.035"_

AMPERES _180 - 190_   VOLTS (arc) _23 - 24_   STICKOUT _1/2" - 3/4"_

SPOTWELD TIME (seconds) ___   BURNBACK TIME (seconds) ___

TRAVEL (forehand) _X_ (backhand) ___ (uphill) ___ (downhill) ___

PRE-WELD CLEANING TYPE _S/STEEL ETCH_

Figure 12-15.

Inverter equipment can provide easy portability for on-site fabrication or maintenance welding. This unit has been paired with a wire feeder for GMA welding a steel frame. (Miller Electric Mfg. Co.)

13 *Truck-Trailer and Off-Road Vehicle Welding Procedures*

Objectives

After studying this chapter, you will be able to:
- Follow procedures for welding truck-trailers and off-road vehicles.
- Inspect and test your own welds.
- Work in compliance with all safety rules.

This chapter details the welding of various joint designs used in making truck-trailers and off-road vehicles. The most common metal used in producing this equipment is steel. It is used in many forms, such as pipe, tubing, angles, channels, beams, plate, forgings, and castings.

Assembly of component parts is a vital step in constructing a truck-trailer. Fit-up of parts is a major variable in producing reliable welds. When parts do not fit together properly, higher stress is placed on the joint. This increased stress can cause weld failure. It is also essential that all parts are clean before being welded. Do not begin welding until you have removed any material that could cause weld defects.

Truck- trailers and off-road vehicles are subject to many areas of stress and vibration while in use. Therefore, only proven weld procedures should be used. The procedures shown in this chapter may require adapting to your welding system. As stated in Chapter 12, always make test welds to verify a new procedure. This helps determine welding parameters and variables for making the best welds. Ultimately, correctness of parameters and variables may be confirmed through destructive testing. Such tests will determine whether proper penetration with good fusion, but without excessive porosity, has been achieved. Quality control procedures must be used to assure that each weld performs its intended function.

Welding Procedures

The weld schedule shown in Figure 13-1 specifies the process mode and all welding parameters used in producing a truck-trailer. All of the welds shown in Figures 13-2 through 13-10 were made with this procedure. With these schedules, all welding should be done in the flat or horizontal positions, if possible. Whenever welding in the vertical or overhead positions, be sure to make a test weld first. This helps predict final quality before actual welding.

After welding, visually inspect each joint for such defects as undercut, overlap, cold shuts, or low crowns. Where appropriate, check fillet weld size. If possible, make a dye penetrant test to identify the presence of cracks. When cracks or defects are found, repair these areas before placing the vehicle into service. Even a small undercut, crack, or other defect can cause a weld to fail when stressed during operation.

GAS METAL ARC WELDING SCHEDULE

PROCESS MODE *SHORT-ARC*

WELD TYPE *FILLET*   POSITION *FLAT/HORIZONTAL*

BASE METAL TYPE *STEEL*   BASE METAL THICKNESS *1/8"*

WIRE TYPE *E70S-4*   WIRE DIA. *.035"*   WIRE SPEED (IPM) *300*

GAS TYPE *Ar-75%*   GAS TYPE *CO_2-25%*   FLOW RATE (CFH) *35*

NOZZLE TYPE *STD*   NOZZLE DIA. *1/2" - .500"*

CONTACT TIP TYPE (long) *X* (short) ___ (standard) ___   DIA. *.035"*

AMPERES *150*   VOLTS (arc) *19*   STICKOUT *3/8"*

SPOTWELD TIME (seconds) ___   BURNBACK TIME (seconds) ___

TRAVEL (forehand) ___ (backhand) ___ (uphill) ___ (downhill) ___

PRE-WELD CLEANING TYPE *WIRE BRUSH*

Figure 13-1. Welding schedule used when welding a steel truck-trailer frame.

Figure 13-2. Overall view of basic truck-trailer framework.

Figure 13-4. Channel-to-channel-to-plate assembly.

Figure 13-3. Channel-to-channel assembly. Note the reinforcing gusset.

Figure 13-5. Channel-to-channel assembly and a steel-rod rope tie.

Figure 13-6. Three-piece section channel assembly.

Figure 13-9. Tubing-to-tubing-to-plate assembly.

Figure 13-7. Channel-to-plate assembly.

Figure 13-10. Tubing-to-tubing-to-plate assembly.

The weld schedule in Figure 13-11 specifies the process mode and all of the parameters for constructing the framework of an off-road vehicle. In use, great stresses are imposed on these vehicles. Therefore, welds must be of a very high grade for the safety of the passengers. All framework welds shown in Figures 13-12 through 13-15 were made with this procedure. Various components are used on the vehicle for attaching struts, springs, shock absorbers, and other items. The weld procedure used for this is found in Figure 13-16. All of the welds in Figures 13-17 through 13-20 were made with this procedure.

Modification and Repair

During a vehicle's lifetime, changes or repairs may be needed to alter the structural design or fix broken welds. A replacement part should fit into place just like the original. If a cracked weld is being repaired, it should

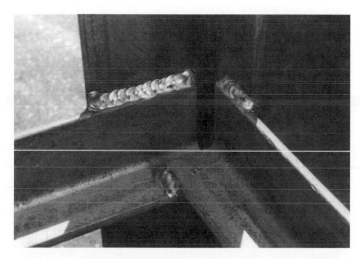

Figure 13-8. Angle-to-angle assembly.

GAS METAL ARC WELDING SCHEDULE

PROCESS MODE *SHORT-ARC*

WELD TYPE *FILLET/GROOVE*   POSITION *FLAT/HORIZONTAL*

BASE METAL TYPE *STEEL*   BASE METAL THICKNESS *12 - 16 ga.*

WIRE TYPE *E70S-4*   WIRE DIA. *.030"*   WIRE SPEED (IPM) *250 - 270*

GAS TYPE *Ar-75%*   GAS TYPE *CO_2-25%*   FLOW RATE (CFH) *35*

NOZZLE TYPE *STD*   NOZZLE DIA. *3/8" - .375"*

CONTACT TIP TYPE (long) *X* (short) ___ (standard) ___ DIA. *.030"*

AMPERES *120 - 130*   VOLTS (arc) *19*   STICKOUT *3/8"*

SPOTWELD TIME (seconds) ___   BURNBACK TIME (seconds) ___

TRAVEL (forehand) ___ (backhand) ___ (uphill) ___ (downhill) ___

PRE-WELD CLEANING TYPE *WIRE BRUSH*

Figure 13-11. Welding schedule used for framework tubing assembly.

Figure 13-12. This typical off-road vehicle required many welds.

Figure 13-14. To mount suspension parts, tubing and plates are usually welded together.

Figure 13-13. The end of the tube was cut to match the outside diameter of the short bearing tube. This type of cut is often referred to as a "fishmouth" cut.

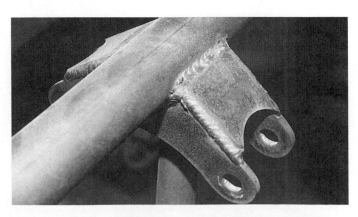

Figure 13-15. Plates are welded together, then assembled to the tube for final welding of the suspension parts.

GAS METAL ARC WELDING SCHEDULE

PROCESS MODE *SHORT-ARC*

WELD TYPE *FILLET*   POSITION *FLAT/HORIZONTAL*

BASE METAL TYPE *STEEL*   BASE METAL THICKNESS *1/8" - 3/16"*

WIRE TYPE *E70S-4*   WIRE DIA. *.035"*   WIRE SPEED (IPM) *290 - 310*

GAS TYPE *Ar-75%*   GAS TYPE *CO_2 -25%*   FLOW RATE (CFH) *35*

NOZZLE TYPE *STD*   NOZZLE DIA. *1/2" - .500"*

CONTACT TIP TYPE (long) *X* (short) ___ (standard) ___ DIA. *.035"*

AMPERES *140 - 150*   VOLTS (arc) *19*   STICKOUT *3/8"*

SPOTWELD TIME (seconds) ___   BURNBACK TIME (seconds) ___

TRAVEL (forehand) ___ (backhand) ___ (uphill) ___ (downhill) ___

PRE-WELD CLEANING TYPE *WIRE BRUSH*

Figure 13-16. Welding schedule used for plate and tubing suspension parts.

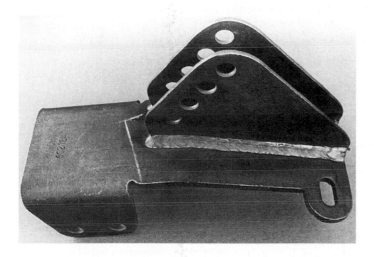

Figure 13-17. Fillet welds were used to weld these preformed parts, which were held in an assembly jig for accurate alignment of the drilled holes.

Figure 13-18. Drilled and preformed plates are attached to the tubing with fillet welds completely around the plate.

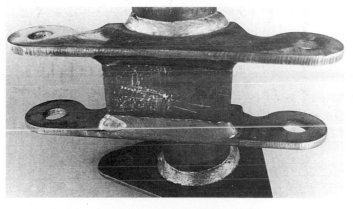

Figure 13-19. Suspension assembly made from tubing and plates with fillet welds. Note the even contour and shape of the welds.

first be slightly grooved so the new weld will penetrate into the root of the joint. All grease, oil, rust, or dirt must be removed by wirebrushing and the part carefully cleaned with a solvent. The weld area should be brought to bright metal condition by wirebrushing or other methods before welding begins.

When welding a tube with closed ends, always drill a small hole into the tube. This will allow hot gases inside the tube to escape. If this is not done, hot gases will

Figure 13-20. The fillet weld size (leg length) is equal to the base metal thickness in this assembly.

expand and exert pressure, blowing away the molten material as you weld. As a result, the tube cannot be closed properly.

Adding more weld metal to a joint or rewelding over an old weld does not work on these vehicles. High weld crown, incomplete penetration, and contamination defects will cause weld failure when placed under stress. Be sure the welds you make are correct and will perform the intended job.

Safety

When working on vehicles containing gas or diesel engines, gas tanks, or batteries, always follow the safety rules outlined in Chapter 14. Take the time to do the job right; injuries can be avoided.

CHAPTER 14 *Autobody Welding Procedures*

Objectives

After studying this chapter, you will be able to:
- Follow safety procedures.
- Prepare an autobody joint for welding.
- Make quality welds using a special autobody welder.
- Remove and replace welded autobody panels.

Modern automobiles are made using lightweight materials. These materials must conform to safety standards and weight limitations (to achieve fuel economy) without sacrificing strength. This is done by using thinner-gauge steels that have been hardened to high tensile strengths before welding. The GMAW process is used with resistance spot welding during manufacturing of many vehicles. For repair of a wrecked vehicle, GMAW has proven to be reliable when applied with proper techniques. GMAW is also used for repair of automobiles that do not use the thin, high-tensile steels.

Various filler materials are used by automobile manufacturers during production. Structural repairs, however, are made with steel filler materials. These include wires in 0.023″, 0.030″, and 0.035″ diameter. All can be used with the GMAW process. Minimum heat input is used to avoid overheating the high-strength base materials. Overheating will temper hardened steel, reducing its tensile strength.

Safety Procedures

Whenever you work on an automobile, remember that you must protect yourself and those around you by following proper safety procedures. You must be alert at all times, since both welding preparation and actual welding operations present many chances for injury. Follow these safety rules:
- **Always wear safety glasses when grinding, drilling, or wirebrushing metal.**
- **Wear proper protective clothing while welding.**
- **Make sure you have the correct protective lens in the welding helmet. (See Chart R-1 in the Reference Section).**
- **Before welding on any part of an automobile, disconnect the battery. If there is any danger of welding sparks falling on or near the battery, remove it or cover it with wet rags.**
- **Clean all spilled fuel from the area before welding. Do not weld near open fuel lines.**
- **If you must weld near a gasoline tank, fill line, or tank vent, make sure all openings are closed to prevent the escape of fumes. Place wet rags, if possible, on areas where sparks may fall.**
- **Always have a fire extinguisher nearby while welding.**
- **When you finish welding, carefully check all areas where sparks have fallen or molten metal has collected. Make sure nothing has been ignited by the sparks or hot metal.**

Cleaning Procedures

Many types of sealing and insulation materials are used in the manufacture of modern automobiles. These materials must be removed from areas where welding is to take place. Such materials (or their residues) may ignite from the sparks or the arc of the welding operation. The fumes are very toxic to the welder and to others in the area. In addition to being toxic, these gases can possibly cause several types of weld defects. These include:
- Porosity.
- Lack of fusion on the weld boundary.
- Cracks in the weld metal.

Dirt, rust, scale, and oil will cause the same problems in welds. For this reason, the area to be welded must be thoroughly cleaned before welding is attempted.

Use wire brushes, files, grinders, or disc sanders to clean the metal surface. **Always wear safety glasses when performing mechanical cleaning operations.** An area that has been cleaned and is ready for welding is shown in Figure 14-1.

Zinc coating (galvanizing) that has been applied to metal to prevent corrosion does not have to be removed before welding. Zinc has a lower melting point than the metals being welded, and rises to the surface of the molten weld pool. Thus, it does not affect weld strength.

Equipment

All types of GMA welding equipment may be used for autobody welding. However, special welding machines have been developed for this purpose, and are very efficient when used for these applications. Smaller machines, which use 110V ac, are very adaptable for light-gauge materials; machines with 208V ac power supplies allow a higher duty cycle for thicker materials.

When the machines are to be used for *spot welds* or *intermittent welds,* they include timers to control the length of the weld cycle, as well as the wire burnback after the weld is completed. Timers may be built into the machine or wire feeder, or they may be added as an additional circuit. Both types of power supplies are manufactured with these options.

The welding machines shown in Figures 14-2 and 14-3 have spot, stitch, and burnback controls for use on autobody welding and other types of weldments. The timers are used to control the length of *on* (arc) time. *Off* time is the time period before the arc starts again, and the *burnback* time is at the end of the cycle to prevent the wire from sticking in the molten pool.

This type of machine includes the welding wire supply and the wire feeder, Figure 14-4. To prevent overspooling when welding is done, adjustments for the

Figure 14-2. This welding power supply has a timer unit installed. (ESAB Welding & Cutting Products)

spool drag must be made in the center of the spool hub. Changing the tension on the spring changes the braking force. The drive rolls are adjusted to control the amount of *pinch* on the wire. If the drive roll pressure is too low, the wire will slip and cause arc outages. If the pressure is too great, the wire may birdnest. These adjustments must be made whenever you change the type or diameter of the welding wire.

Welding machines must operate under the adverse conditions usually found in repair shops. Therefore, they require special maintenance procedures for continued proper operation. It is strongly advised that you follow the operating manual recommendations to obtain the best results from the machine.

Welding Guns

The guns used for autobody GMA welding are usually gas-cooled, since the operating amperages are low and the time periods for welding are short. Usually, guns with a 100-ampere capacity and a 20% to 40% duty cycle will be sufficient for autobody work. Too large a gun often will be difficult to handle in cramped autobody work areas. A typical GMAW gun is shown in Figure 14-5.

Many spot and corner-type joints require welding with specially designed nozzles. Spot and corner-type nozzles are shown in Chapter 8.

Special Tools

Plug-type welds require that a hole be made in the upper side of the mating materials. This is usually done with a hole punch that produces a 1/4″ or 5/16″ hole. The tool shown in Figure 14-6 is hand-operated, while the tool shown in Figure 14-7 is operated by air pressure.

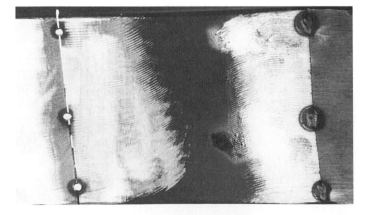

Figure 14-1. Proper cleaning is important to making good welds. The painted area has been cleaned with a polishing disc and tackwelded to hold the pieces in correct position.

Figure 14-3. This 208V ac GMA welding unit includes a digital voltage control for precise setting of the arc gap. (Lincoln Electric Co.)

Figure 14-4. Adjustments for both wire supply reel tension and drive roll pressure are made inside the wire supply cabinet. (CK Worldwide, Inc.)

Joggle-type joints require the lower part of the assembly be flanged so the mated parts will have a flat upper surface. The flange may be bent by using the hole-punching tool, or it may be done with a single-purpose tool like the one shown in Figure 14-8. An assembled joggle joint is shown in Figure 14-9.

Welding Procedures

The welding procedure to be used is governed by a number of factors. These may include:

- Type of joint.
- Type of weld.
- Base material thickness.
- Size (diameter) of the filler wire.
- Type of shielding gas.

The manufacturer of the welding machine will generally have developed and tested welding procedures for several different types of welds. These procedures are specific for this machine. Figure 14-10 is a set-up chart

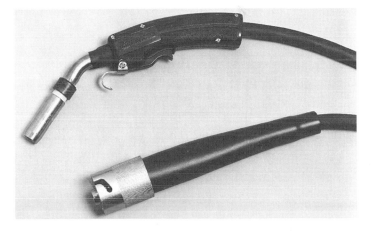

Figure 14-5. Gas-cooled welding gun used for GMAW applications involving relatively low amperage and short welding cycle times. (Bernard Welding Equipment Co.)

for a welding machine. Such a chart may be found on the door of the welding machine, or in the operating manual for the machine.

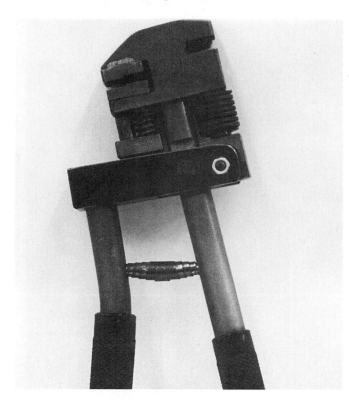

Figure 14-6. This hand-operated tool can punch holes for plug welds or bend flanges for joggle joints.

Figure 14-8. A manual tool used to bend flanges for joggle joints.

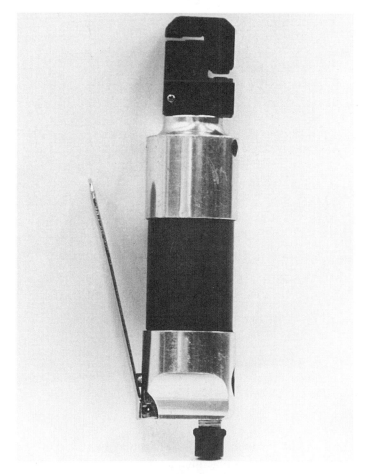

Figure 14-7. An air-operated (pneumatic) hole puncher is useful when many holes must be made.

The displayed machine settings are general in nature. To define the actual parameters for the weld, a test should be made using the factors listed on page 133. Visual and/or destructive testing may then be carried out to prove the quality of the weld procedure. A simple test weldment for plug welds is shown in Figure 14-11.

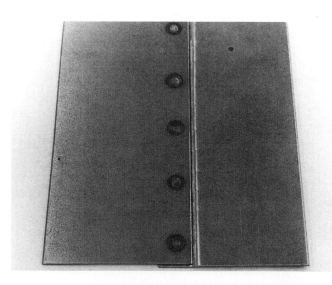

Figure 14-9. A joggle joint is designed to provide a flat surface. In this example, resistance spot welds hold the assembly together for final welding. Tack welds or clamps are used for the same purpose.

MM-251 Weld Setup Guide

Follow the steps below to set welding condition in your Migmaster 251:

1. Find the chart for the type of base plate material and wire type to be welded.
2. Find the thickness for the plate that will be welded.
3. Follow the column that matches the plate thickness down to the row that matches the welding wire diameter being used. This is the recommended VOLTAGE (v) and WIRE FEED (ipm) settings.
4. Set the front panel dials to match the recommended settings from the chart. The top dial is WIRE FEED (ipm) and the bottom dial is VOLTAGE (v).
5. To fine-tune welding performance, increase or decrease the VOLTAGE dial. NOTE! In most cases, it will be unnecessary to change setting more that 1 volt higher or lower than the recommended setting.

Carbon Steel Base Plate with ESAB Spoolarc 87HP Solid Wire

Shielding Gas — Short Circuiting Mig (GMAW): 75% Argon - 25% CO_2 (for 100% CO_2 add 1 volt to recommend volt setting) | Spray Arc: 98% Argon - 2% CO_2

Wire Diameter	24 GA	20 GA	18 GA	16 GA	14 GA	12 GA	10 GA	3/16"	1/4"+	Spray 3/16"	Spray 1/4"+
.023"		150 ipm / 14.5 v	175 ipm / 15.0 v	200 ipm / 15.5 v	250 ipm / 16.0 v	275 ipm / 16.5 v	300 ipm / 17.0 v				
.030"	100 ipm / 14.5 v	125 ipm / 15.0 v	150 ipm / 15.5 v	175 ipm / 15.5 v	200 ipm / 16.0 v	250 ipm / 17.0 v	275 ipm / 17.5 v				
.035"			125 ipm / 15.5 v	150 ipm / 15.5 v	175 ipm / 16.0 v	200 ipm / 17.0 v	250 ipm / 17.5 v	350 ipm / 18.0 v		350 ipm / 24.5 v	350 ipm / 25.0 v
.045"				75 ipm / 16.5 v	100 ipm / 17.0 v	140 ipm / 17.5 v	150 ipm / 18.0 v	175 ipm / 18.5 v	250 ipm / 19.0 v	250 ipm / 25.0 v	250 ipm / 25.0 v

Carbon Steel Base Plate with ESAB Dual Shield 7100 Cored Wire

Shielding Gas: 75% Argon - 25% CO_2 (for 100% CO_2 add 1 volt to recommend volt setting)

Wire Diameter	24 GA	20 GA	18 GA	16 GA	14 GA	12 GA	10 GA	3/16"	1/4"+
.035"							250 ipm / 23.5 v	275 ipm / 25.0 v	350 ipm / 26.5 v
.045"							175 ipm / 24.0 v	200 ipm / 25.5 v	250 ipm / 27.0 v

Stainless Steel Base Plate with ESAB Arcaloy ER308L Solid Wire

Shielding Gas: Tri-Mix (7.5% Argon - 2.5% CO_2 - 90% Helium)

Wire Diameter	24 GA	20 GA	18 GA	16 GA	14 GA	12 GA	10 GA	3/16"	1/4"+
.035"				125 ipm / 18.0 v	125 ipm / 18.0 v	225 ipm / 18.0 v	250 ipm / 19.0 v	300 ipm / 20.0 v	300 ipm / 21.0 v
.045"					125 ipm / 17.0 v	150 ipm / 17.5 v	175 ipm / 18.0 v	200 ipm / 18.5 v	225 ipm /

Aluminum Base Plate with ESAB 5356HQ Solid Wire

Shielding Gas: 100% Argon

Wire Dia.	24 GA	20 GA	18 GA	16 GA	14 GA	12 GA	10 GA	3/16"	1/4"+
3/64"						275 ipm / 20.0 v	300 ipm / 21.0 v	325 ipm / 22.5 v	400 ipm / 24.5 v

Carbon Steel Base Plate with ESAB Coreshield 15 Gasless Wire

Shielding Gas: No Shielding Gas Required

Wire Diameter	24 GA	20 GA	18 GA	16 GA	14 GA	12 GA	10 GA	3/16"	1/4"+
.030"			100 ipm / 14.0 v (min.)	125 ipm / 15.0 v	150 ipm / 17.5 v	200 ipm / 18.0 v	300 ipm / 19.0 v	350 ipm / 20.0 v	400 ipm / 20.5 v
.035"				100 ipm / 16.0 v	125 ipm / 17.0 v	175 ipm / 18.5 v	200 ipm / 19.0 v	250 ipm / 19.0 v	300 ipm / 20.0 v
.045"					100 ipm / 16.0 v	125 ipm / 18.0 v	150 ipm / 19.0 v	175 ipm / 19.0 v	/ 20.5 v

Figure 14-10. The set-up chart for a welding machine lists the actual welding voltage and wire speed for various types and thicknesses of metals with different types of shielding gases. (ESAB Welding & Cutting Products)

Whenever any of the factors listed changes or the quality of the weld changes, additional testing should be done to verify weld quality.

To be able to *repeat* a satisfactory procedure, record the actual welding parameters on a weld schedule, using measured wire feed in inches per minute. Set the arc voltage gap settings as close as possible with a defined stickout from the end of the contact tip or the gas nozzle.

Unibody Repair Procedures

Many modern automobiles are assembled with resistance spot or seam welds made by robots. These welds must be removed to take the damaged panels off the vehicle. This is done by grinding the weld from the upper panel, or by using routers or drills ground to a flat angle. Several types of routers (rotary files) are shown in Figure 14-12. If additional welds are needed in the replacement panel, holes may be punched or drilled as required. The manufacturer's recommendation for hole

spacing should be followed for proper strength in the joint area. Where this dimension is not known, a 1″ to 1 1/2″ spacing can be used. The joint area is then cleaned, and the new components installed. Figure 14-13 and Figure 14-14 show replacement panels held in place with clamps to properly position them for welding.

Where clamps cannot be used, drill small holes in the aligned panels where the weld is to be located. Then,

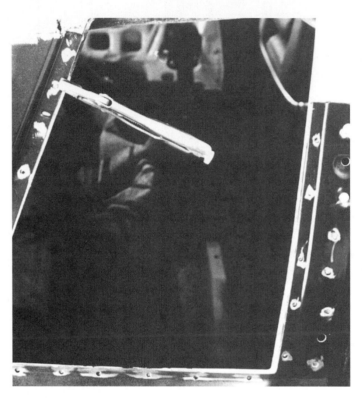

Figure 14-13. Cleaning should be completed before assembling replacement panels to the automobile.

Figure 14-11. In this test weld, the material thickness and joint design of the desired weld have been used. Visual and destructive tests may be performed to confirm weld quality.

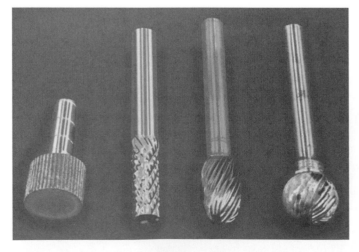

Figure 14-12. Router bits like these are used to cut through original spot welds when removing body panels.

Figure 14-14. Clamps are often used to align and secure replacement panels for welding.

install small sheet metal screws to hold the assembly together. Tack weld the panels, remove the screws, then drill out each upper panel hole for a plug weld.

To assure good starting of the arc, always cut the end of the weld wire to remove oxidation and achieve the desired stickout. See Figure 14-15. This is especially helpful when making plug or spot welds, where excessive wire stickout might cause cold shuts or lack of fusion. Figures 14-16 and 14-17 show butt, fillet, plug, and spot welds used to complete the repair of a vehicle.

Sample Welding Procedures

The welding procedures shown in Figures 14-18 through 14-22 are general in nature and should be qualified before use.

Figure 14-16. These welds are spaced more closely than those made at the factory during assembly.

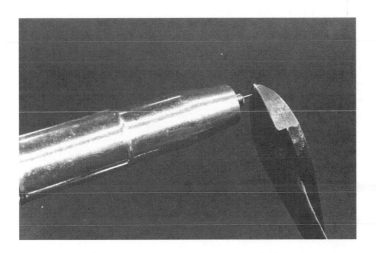

Figure 14-15. Oxidized wire and excessive stickout should be clipped off before attempting to weld.

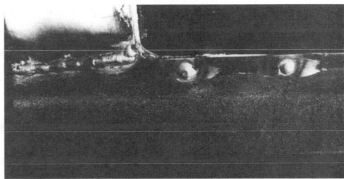

Figure 14-17. Once welds are completed, they must be ground and polished before finishes can be applied.

GAS METAL ARC WELDING SCHEDULE

PROCESS MODE *SHORT-ARC*

WELD TYPE *PLUG*   POSITION *FLAT*

BASE METAL TYPE *STEEL*   BASE METAL THICKNESS *.030" TOP*

WIRE TYPE *E70S-6*   WIRE DIA. *.030"*   WIRE SPEED (IPM) *125 - 135*

GAS TYPE *Ar - 75%*   GAS TYPE *CO_2 - 25%*   FLOW RATE (CFH) *35*

NOZZLE TYPE *STD*   NOZZLE DIA. *1/2" - .500"*

CONTACT TIP TYPE (long) *X*   (short) ___   (standard) ___   DIA. *.030"*

AMPERES _____   VOLTS (arc) *15 - 16*   STICKOUT *1/4"*

SPOTWELD TIME (seconds) ___   BURNBACK TIME (seconds) ___

TRAVEL (forehand) ___   (backhand) ___   (uphill) ___   (downhill) ___

PRE-WELD CLEANING TYPE *DEGREASE*

A

B

Figure 14-18. Lap joint plug weld. A—Weld crown (top of plate). B—Penetration (bottom of plate). The penetration indicates these are acceptable welds.

GAS METAL ARC WELDING SCHEDULE

PROCESS MODE *SHORT-ARC*

WELD TYPE *FILLET*   POSITION   *FLAT*

BASE METAL TYPE *STEEL*   BASE METAL THICKNESS   *.035" TOP*

WIRE TYPE *E70S -6*   WIRE DIA. *.030"*   WIRE SPEED (IPM) *135 - 145*

GAS TYPE *Ar - 75%*   GAS TYPE   *CO$_2$ - 25%*   FLOW RATE (CFH) *30*

NOZZLE TYPE *STD*   NOZZLE DIA. *1/2" - .500"*

CONTACT TIP TYPE (long) *X*   (short) ___   (standard) ___   DIA. *.030"*

AMPERES ____   VOLTS (arc) *16 - 17*   STICKOUT *1/4"*

SPOTWELD TIME (seconds) ___   BURNBACK TIME (seconds) ___

TRAVEL (forehand) *X*   (backhand) ___   (uphill) ___   (downhill) ___

PRE-WELD CLEANING TYPE *DEGREASE*

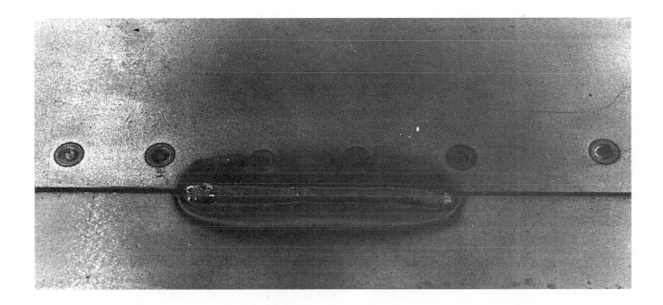

Figure 14-19. Lap joint filler weld.

GAS METAL ARC WELDING SCHEDULE

PROCESS MODE *SHORT-ARC*

WELD TYPE *SPOT*   POSITION *HORIZONTAL*

BASE METAL TYPE *STEEL*   BASE METAL THICKNESS *.059"*

WIRE TYPE *E70S -6*   WIRE DIA. *.030"*   WIRE SPEED (IPM) *125 - 135*

GAS TYPE *Ar - 75%*   GAS TYPE *CO_2 - 25%*   FLOW RATE (CFH) *30*

NOZZLE TYPE *SPOT- SLOTTED*   NOZZLE DIA. *5/8" - .620"*

CONTACT TIP TYPE (long) ___ (short) *X* (standard) ___   DIA. ____

AMPERES ____   VOLTS (arc) ____   STICKOUT *1/8" -1/4"*

SPOTWELD TIME (seconds) *4*   BURNBACK TIME (seconds) *1/4*

TRAVEL (forehand) ___ (backhand) ___ (uphill) ___ (downhill) ___

PRE-WELD CLEANING TYPE *SAND–GRIND–DEGREASE*

Figure 14-20. Lap joint spot weld.

GAS METAL ARC WELDING SCHEDULE

PROCESS MODE *SHORT-ARC*

WELD TYPE *BUTT*   POSITION   *FLAT*

BASE METAL TYPE *STEEL*   BASE METAL THICKNESS   *.030"*

WIRE TYPE *E70S-4*   WIRE DIA. *.023"*   WIRE SPEED (IPM) *160"*

GAS TYPE *Ar-75%*   GAS TYPE   *CO_2-25%*   FLOW RATE (CFH) *30*

NOZZLE TYPE *STD*   NOZZLE DIA. *3/8"-.375"*

CONTACT TIP TYPE (long) *X*   (short) ___   (standard) ___   DIA. *.023"*

AMPERES _____   VOLTS (arc) _____   STICKOUT *1/4"*

SPOTWELD TIME (seconds) ___   BURNBACK TIME (seconds) ___

TRAVEL (forehand) ___   (backhand) ___   (uphill) ___   (downhill) ___

PRE-WELD CLEANING TYPE *GRIND–SAND–WIRE BRUSH*

Figure 14-21. Butt joint tack welds. The grinder is used to remove the crowns of the tack welds before the longseam weld is made. Removing the crowns aids penetration.

GAS METAL ARC WELDING SCHEDULE

PROCESS MODE *SHORT-ARC*

WELD TYPE *JOGGLE*   POSITION   *FLAT*

BASE METAL TYPE *STEEL*   BASE METAL THICKNESS *.024" - 030"*

WIRE TYPE *E70S - 4*   WIRE DIA. *.023"*   WIRE SPEED (IPM) *180"*

GAS TYPE *Ar - 75%*   GAS TYPE *CO$_2$ - 25%*   FLOW RATE (CFH) *30*

NOZZLE TYPE *STD*   NOZZLE DIA. *1/2" - .500"*

CONTACT TIP TYPE (long) *X* (short) ___ (standard) ___ DIA. *.023"*

AMPERES ___   VOLTS (arc) ___   STICKOUT *3/8"*

SPOTWELD TIME (seconds) ___   BURNBACK TIME (seconds) ___

TRAVEL (forehand) *X* (backhand) ___ (uphill) ___ (downhill) ___

PRE-WELD CLEANING TYPE *DEGREASE*

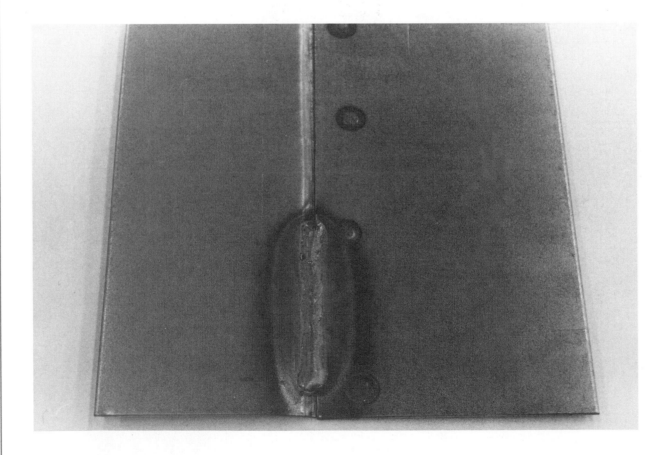

Figure 14-22. Joggle joint seam weld.

Review Questions

Write your answers on a separate sheet of paper. Do not write in this book.

1. Compare modern steels used in automaking with steels used in the past. How are they different?
2. Aside from GMAW, what other welding process is commonly used in automobile manufacturing?
3. Using GMAW for vehicle repair requires proper techniques for _____ welds.
4. What welding wire diameters are usually used when doing structural repairs on unibody vehicles?
5. When too much _____ is used during the welding of high-strength steels, the material loses its tensile strength.
6. Improper removal of dirt, paint, insulation, or other foreign matter from the joint area before welding will result in _____. _____, and _____ may also occur.
7. Must a zinc coating applied for corrosion protection be removed before metal is welded?
8. The melting point of zinc is _____ than that of steel.
9. Why do welding machines used in autobody repair require careful attention to maintenance?
10. Holes punched in metal for plug welding are typically _____″ or _____″ in diameter.
11. Joggle joints are made by _____ the lower piece of the assembly so that the top surface of the joint will be _____.
12. Hole punches may be hand-powered or operated by _____ _____.
13. What are the two main adjustments on an autobody welding machine?
14. To ensure quality results, what should always be done before beginning any procedure with the GMAW process?
15. To dismount damaged panels from a vehicle, welds are removed by _____ or the use of routers or _____-_____ drills.
16. If oxidized wire and excess stickout are not cut away before starting welding, the weld may show _____ _____ or _____ _____ _____.

Safety Test

Write your answers on a separate sheet of paper. Do not write in this book.

1. You should always wear _____ _____ when welding, grinding, drilling, or wirebrushing.
2. How does a smart welder avoid burns from sparks or hot metal?
3. How do you make sure your eyes are properly protected from arc radiation?
4. What safety precaution should be taken before beginning to use any arc welding process on an automobile?
5. If sparks are likely to fall on or near the vehicle battery, cover it with _____ _____.
6. Always clean up any spilled _____ from an area before welding.
7. When welding near gasoline tanks, vents, or fill openings, make sure the openings are _____ and covered with wet rags.
8. _____ _____ should be placed over upholstery if sparks are likely to fall on it.
9. Before starting to weld, make sure there is a fire _____ available nearby.
10. After finishing welding, check all areas where _____ or _____ _____ may have collected, to avoid the possibility of fire.

Welding students develop GMA skills through hands-on experience under the close guidance of their instructors. This instructor is checking an example of the student's work. (American Welding Society)

Reference Section

The following pages contain more than 20 charts that will be useful as reference in a variety of welding-related areas. To make locating information easier, the charts are listed below by title and page number.

Guide for Shade Numbers			
Operation	**Electrode Size 1/32 in. (mm)**	**Arc Current (A)**	**Minimum Protective Shade**	**Suggested[1] Shade No. (Comfort)**

Operation	Electrode Size 1/32 in. (mm)	Arc Current (A)	Minimum Protective Shade	Suggested[1] Shade No. (Comfort)
Shielded metal arc welding	Less than 3 (2.5) 3–5 (2.5–4) 5–8 (4–6.4) More than 8 (6.4)	Less than 60 60–160 160–250 250–550	7 8 10 11	— 10 12 14
Gas metal arc welding and flux cored arc welding		Less than 60 60–160 160–250 250–500	7 10 10 10	— 11 12 14
Gas tungsten arc welding		Less than 50 50–150 150–500	8 8 10	10 12 14
Air carbon arc cutting	(Light) (Heavy)	Less than 500 500–1000	10 11	12 14
Plasma arc welding		Less than 20 20–100 100–400 400–800	6 8 10 11	6 to 8 10 12 14
Plasma arc cutting	(Light)[2] (Medium)[2] (Heavy)[2]	Less than 300 300–400 400–800	8 9 10	9 12 14
Torch brazing		—	—	3 or 4
Torch soldering		—	—	2
Carbon arc welding		—	—	14

	Plate thickness			
	in.	mm		
Gas welding Light Medium Heavy	Under 1/8 1/8 to 1/2 Over 1/2	Under 3.2 3.2 to 12.7 Over 12.7		4 or 5 5 or 6 6 or 8
Oxygen cutting Light Medium Heavy	Under 1 1 to 6 Over 6	Under 25 25 to 150 Over 150		3 or 4 4 or 5 5 or 6

[1] As a rule of thumb, start with a shade that is too dark to see the weld zone. Then go to a lighter shade which gives sufficient view of the weld zone without going below the minimum. In oxyfuel gas welding or cutting where the torch produces a high yellow light, it is desirable to use a filter lens that absorbs the yellow or sodium line in the visible light of the (spectrum) operation.

[2] These values apply where the actual arc is clearly seen. Experience has shown that lighter filters may be used when the arc is hidden by the workpiece.

Chart R-1. Guide for Shade Numbers.

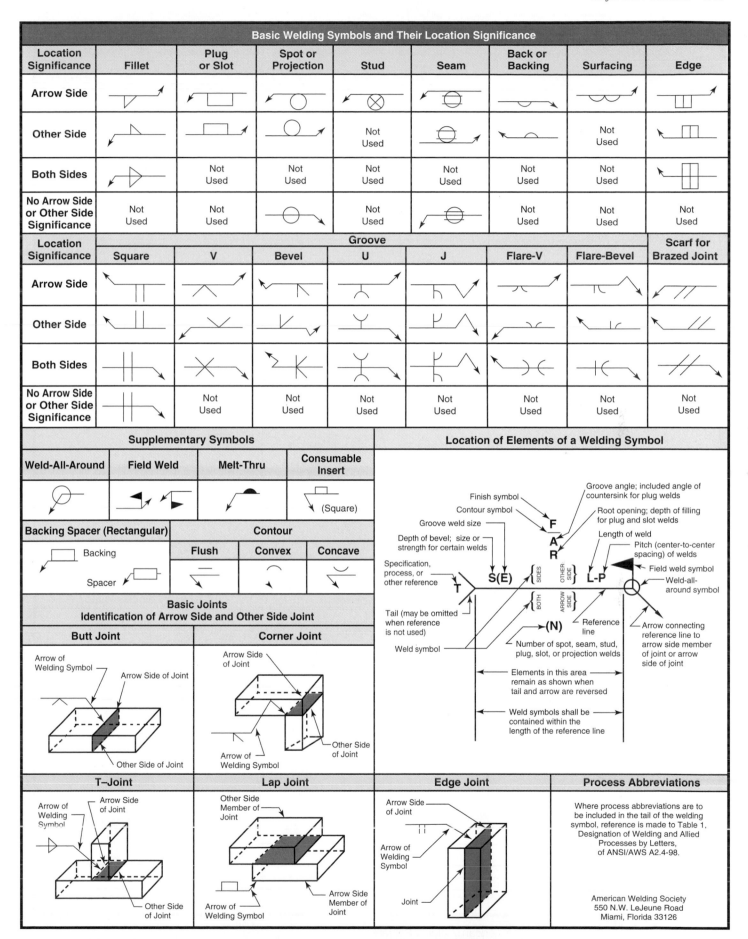

Chart R-2. Basic Welding Symbols and Their Location Significance. (American Welding Society)

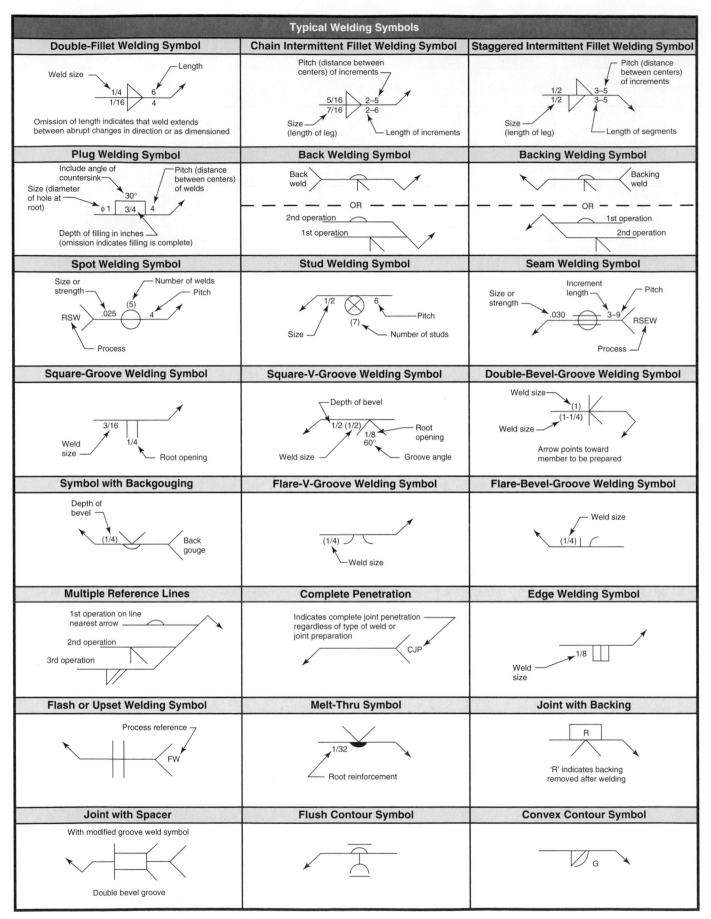

Chart R-3. Typical Welding Symbols. (American Welding Society)

Chemical Treatments for Removal of Oxide Films from Aluminum Surfaces			
Solution	**Concentration**	**Procedure**	**Purpose**
Nitric Acid	50% water, 50% nitric acid, technical grade.	Immersion 15 min. Rinse in cold water, then in hot water. Dry.	Removal of thin oxide film for fusion welding.
Sodium hydroxide (caustic soda) followed by	5% Sodium hydroxide in water.	Immersion 10-60 seconds. Rinse in cold water.	Removal of thick oxide film for all welding processes.
Nitric acid	Concentrated	Immerse for 30 seconds. Rinse in cold water, then hot water. Dry.	Removal of thick oxide film for all welding processes.
Sulfuric-chromic	H_2SO_4 1 gal. CrO_3 45 oz. Water 9 gal.	Dip for 2-3 min. Rinse in cold water, then hot water. Dry.	Removal of films and stains from heat treating, and oxide coatings.
Phosphoric-chromic	H_3PO_3 (75%) 3.5 gal. CrO_3 1.75 lbs. Water 10 gal.	Dip for 5-10 min. Rinse in cold water. Rinse in hot water. Dry.	Removal of anodic coatings.

Chart R-4. Chemical Treatments for Removal of Oxide Films from Aluminum Surfaces.

Inches per Pound of Wire												
Wire Diameter												
Decimal	**Fraction**	**Mag.**	**Alum.**	**Alum. Bronze (10)%**	**Stain-less Steel**	**Mild Steel**	**Stain-less Steel**	**Si. Bronze**	**Copper Nickel**	**Nickel**	**De-ox. Copper**	**Ti**
.020		50500	32400	11600	11350	11100	10950	10300	9950	9900	9800	19524
.025		34700	22300	7960	7820	7680	7550	7100	6850	6820	6750	12492
.030		22400	14420	5150	5050	4960	4880	4600	4430	4400	4360	8776
.035		16500	10600	3780	3720	3650	3590	3380	3260	3240	3200	6372
.040		12600	8120	2900	2840	2790	2750	2580	2490	2480	2450	4884
.045	3/64	9990	6410	2290	2240	2210	2210	2040	1970	1960	1940	3852
.062	1/16	5270	3382	1220	1180	1160	1140	1070	1040	1030	1020	2028
.078	5/64	3300	2120	756	742	730	718	675	650	647	640	
.093	3/32	2350	1510	538	528	519	510	480	462	460	455	964.8
.125	1/8	1280	825	295	289	284	279	263	253	252	249	499.92

Chart R-5. Inches per Pound of Wire.

Weld Metal Requirements for Fillet Welds				
Size of Fillet	45 Deg. Fillets		30-60 Deg. Fillets	
Inches	Lbs. of Metal per Foot	Lbs. of Rod per Foot	Lbs. of Metal per Foot	Lbs. of Rod per Foot
1/8	.027	.039	.054	.078
3/16	.063	.090	.126	.180
1/4	.106	.151	.212	.302
5/16	.166	.237	.332	.474
3/8	.239	.342	.478	.684
7/16	.325	.465	.650	.930
1/2	.425	.607	.850	1.214
5/8	.663	.948	1.226	1.896
3/4	.955	1.364	1.910	2.728
7/8	1.300	1.857	2.600	3.714
1	1.698	2.425	3.396	4.850

Chart R-6. Weld Metal Requirements for Fillet Welds.

Current Ranges for GMAW of Steel with Short-Arc Transfer				
Welding Current				
Wire Diameter	Flat Position		Vertical and Overhead Positions	
	Minimum	Maximum	Minimum	Maximum
.030 in.	50	150	50	125
.035 in.	75	175	75	150
.045 in.	100	225	100	175

Chart R-7. Current Ranges for GMAW of Steel with Short-Arc Transfer.

Spray-Arc Transition Currents for Electrode Types			
Electrode Type	Dia.	Shielding Gas	Minimum Current (Amperes)
Mild Steel	.030	98% Ar-2% Oxygen	150
Mild Steel	.035	98% Ar-2% Oxygen	165
Mild Steel	.045	98% Ar-2% Oxygen	220
Mild Steel	.062	98% Ar-2% Oxygen	275
Stainless Steel	.035	99% Ar-1% Oxygen	170
Stainless Steel	.045	99% Ar-1% Oxygen	225
Stainless Steel	.062	99% Ar-1% Oxygen	285
Aluminum	.030	Argon	95
Aluminum	.045	Argon	135
Aluminum	.062	Argon	180
Deoxidized Copper	.035	Argon	180
Deoxidized Copper	.045	Argon	210
Deoxidized Copper	.062	Argon	310
Silicon Bronze	.035	Argon	165
Silicon Bronze	.045	Argon	205
Silicon Bronze	.062	Argon	270

Chart R-8. Spray-Arc Transition Currents for Electrode Types.

Typical Arc Voltages for Short-Arc Transfer with .035 Inch Diameter Wire			
Metal	Argon	75% Argon/ 25% Carbon Dioxide	Carbon Dioxide
Aluminum	19	—	—
Magnesium	16	—	—
Carbon Steel	—	19	20
Low Alloy Steel	—	19	20
Stainless Steel	—	21	—
Copper	24	—	—
Copper Nickel	23	—	—
Silicon Bronze	23	—	—
Aluminum Bronze	23	—	—
Phosphor Bronze	23	—	—

Chart R-9. Typical Arc Voltages for Short-Arc Transfer with .035″ Wire.

Chemical Compositions of Mild Steel and Low-Alloy Steel Wires							
AWS Class	Carbon	Mang.	Silicon	Sulphur	Phos.	Molyb.	Other
E70S-1	.07-.19	.90-1.40	.30-.50	.035	.025	—	—
E70S-2	.06	.90-1.40	.40-.70	.035	.025	—	.05-.15 Ti. .02-.12 Zr. .05-.15 Al.
E70S-3	.06-.15	.90-1.40	.45-.70	.035	.025	—	—
E70S-4	.07-.15	.90-1.40	.65-.85	.035	.025	—	—
E70S-5	.07-.19	.90-1.40	.30-.60	.035	.025	—	.50-.90 Al.
E70S-6	.07-.15	1.40-1.85	.80-1.15	.035	.025	—	—
E70S-1B	.07-.12	1.60-2.10	.50-.80	.035	.025	.40/.60	—

Chart R-10. Chemical Compositions of Mild Steel and Low-Alloy Steel Wires.

GMAW Process Parameters				
Steel Material		.030 in. Diameter Wire	Short-Arc Mode	
Thickness	Gas	Amperes	Wire Speed	Volts
22 ga.	75 Ar-25 CO_2	40-55	90-100	15-16
20 ga.	75 Ar-25 CO_2	50-60	120-135	15-16
18 ga.	75 Ar-25 CO_2	70-80	150-175	16-17
16 ga.	75 Ar-25 CO_2	90-110	220-250	17-18
14 ga.	75 Ar-25 CO_2	120-130	250-340	17-18

Chart R-11a. GMAW Process Parameters, Short-Arc Mode.

GMAW Process Parameters				
Steel Material		.035 in. Diameter Wire	Short-Arc Mode	
Thickness	Gas	Amperes	Wire Speed	Volts
1/8 in.	75 Ar-25 CO_2	140-150	280-300	18-19
3/16 in.	75 Ar-25 CO_2	160-170	320-340	18-19
1/4 in.	75 Ar-25 CO_2	180-190	360-380	21-22
5/16 in.	75 Ar-25 CO_2	200-210	400-420	21-22
3/8 in.	75 Ar-25 CO_2	220-250	420-520	23-24

Chart R-11b.

GMAW Process Parameters				
Steel Material		.030 in. Diameter Wire	Short-Arc Mode	
Thickness	Gas	Amperes	Wire Speed	Volts
22 ga.	CO_2	40-55	90-100	16-17
20 ga.	CO_2	50-60	120-135	17-18
18 ga.	CO_2	70-80	150-175	18-19
16 ga.	CO_2	90-110	220-250	19-20
14 ga.	CO_2	120-130	250-340	20-21

Chart R-11c.

GMAW Process Parameters				
Steel Material		.035 in. Diameter Wire	Short-Arc Mode	
Thickness	Gas	Amperes	Wire Speed	Volts
1/8 in.	CO_2	140-150	280-300	21-22
3/16 in.	CO_2	160-170	320-340	21-22
1/4 in.	CO_2	180-190	360-380	23-24
5/16 in.	CO_2	200-210	400-420	23-24

Chart R-11d.

GMAW Process Parameters				
Steel Material		.045 in. Diameter Wire	Short-Arc Mode	
Thickness	Gas	Amperes	Wire Speed	Volts
1/8 in.	75 Ar-25 CO_2	140-150	140-150	18-19
3/16 in.	75 Ar-25 CO_2	160-170	160-175	18-19
1/4 in.	75 Ar-25 CO_2	180-190	185-195	21-22
5/16 in.	75 Ar-25 CO_2	200-210	210-220	21-22
3/8 in.	75 Ar-25 CO_2	220-250	220-270	23-24

Chart R-11e.

GMAW Process Parameters				
Stainless Steel Material		.035 in. Diameter Wire	Short-Arc Mode	
Thickness	Gas	Amperes	Wire Speed	Volts
18 ga.	Tri-Mix	50-60	120-150	19-20
16 ga.	Tri-Mix	70-80	180-205	19-20
14 ga.	Tri-Mix	90-110	230-275	20-21
12 ga.	Tri-Mix	120-130	300-325	20-21
3/16 in.	Tri-Mix	140-150	350-375	21-22
1/4 in.	Tri-Mix	160-170	400-425	21-22
5/16 in.	Tri-Mix	180-190	450-475	21-22

Chart R-11f.

GMAW Process Parameters				
Steel Material		.035 in. Diameter Wire	Spray-Arc Mode	
Thickness	Gas	Amperes	Wire Speed	Volts
1/8 in.	98 Ar-2 O_2	160-170	320-340	23-24
3/16 in.	98 Ar-2 O_2	180-190	360-380	24-25
1/4 in.	98 Ar-2 O_2	200-210	400-420	24-25
5/16 in.	98 Ar-2 O_2	220-250	420-520	25-26

Chart R-12a. GMAW Process Parameters, Spray-Arc Mode.

GMAW Process Parameters				
Steel Material		.045 in. Diameter Wire	Spray-Arc Mode	
Thickness	Gas	Amperes	Wire Speed	Volts
1/8 in.	98 Ar-2 O_2	160-170	160-175	23-24
3/16 in.	98 Ar-2 O_2	180-190	185-195	24-25
1/4 in.	98 Ar-2 O_2	200-210	210-220	24-25
5/16 in.	98 Ar-2 O_2	220-250	220-270	25-26
3/8 in.	98 Ar-2 O_2	300 up	375 up	26-27
1/2 in.	98 Ar-2 O_2	315 up	390 up	29-30

Chart R-12b.

GMAW Process Parameters				
Stainless Steel Material		.035 in. Diameter Wire	Spray-Arc Mode	
Thickness	Gas	Amperes	Wire Speed	Volts
3/16 in.	98 Ar-2 O_2	160-170	400-425	23-24
1/4 in.	Tri-Mix	180-190	450-475	24-25

Chart R-12c.

GMAW Process Parameters				
Stainless Steel Material		1/16 in. Diameter Wire	Spray-Arc Mode	
Thickness	Gas	Amperes	Wire Speed	Volts
3/8 in.	98 Ar-2 O_2	220-250	As Req'd	25-26
7/16 in.	98 Ar-2 O_2	300 up	As Req'd	26-27
1/2 in.	98 Ar-2 O_2	325 up	As Req'd	27-32

Chart R-12d.

GMAW Process Parameters				
Aluminum Material		.035 in. Diameter Wire	Spray-Arc Mode	
Thickness	Gas	Amperes	Wire Speed	Volts
1/8 in.	Argon	110-130	350-400	21-22
3/16 in.	Argon	140-150	425-450	23-24

Chart R-12e.

| GMAW Process Parameters | | | | |
| Aluminum Material | 3/64 in. Diameter Wire | | Spray-Arc Mode | |
Thickness	Gas	Amperes	Wire Speed	Volts
3/16 in.	Argon	140-150	300-325	23-24
1/4 in.	Argon	180-210	350-375	24-25
5/16 in.	Argon	200-230	400-425	26-27
3/8 in.	Argon	220-250	450-480	26-28

Chart R-12f.

| GMAW Process Parameters | | | | |
| Aluminum Material | 1/16 in. Diameter Wire | | Spray-Arc Mode | |
Thickness	Gas	Amperes	Wire Speed	Volts
1/4 in.	Argon	180-210	170-185	24-25
5/16 in.	Argon	200-230	200-210	26-27
3/8 in.	Argon	220-250	220-230	26-28
7/16 in.	Argon	280 up	240-270	28-29
1/2 in.	Argon	300 up	290-300	29-30

Chart R-12g.

| GMAW Process Parameters | | | | |
| Aluminum Material | .030 in. Diameter Wire | | Spray-Arc Mode | |
Thickness	Gas	Amperes	Wire Speed	Volts
.062 in.	Argon	90	365	18
.125 in.	Argon	125	440	20

Chart R-12h.

| Etching Reagents for Microscopic Examination of Iron and Steel | | | |
Application	Etching	Composition	Remarks
Carbon low-alloy and medium-alloy steels	Nital	Nitric acid (sp gr 1.42) .... 1–5 ml Ethyl or methyl alcohol .. 95–99 ml	Darkens perlite and gives contrast between adjacent colonies; reveals ferrite boundaries; differentiates ferrite from martensite; shows case depth of nitrided steel. Etching time: 5–60 secs.
	Picral	Picric acid .... 4 g Methyl alcohol .... 100 ml	Used for annealed and quench-hardened carbon and alloy steel. Not as effective as No. 1 for revealing ferrite grain boundaries. Etching time: 5–120 secs.
	Hydrochloric and picral acids	Hydrochloric acid .... 5 ml Picric acid .... 1 g Methyl alcohol .... 100 ml	Reveals austenitic grain size in both quenched and quenched-tempered-steels.
Alloy and stainless steel	Mixed acids	Nitric acid .... 10 ml Hydrochloric acid .... 20 ml Glycerol .... 20 ml Hydrogen peroxide .... 10 ml	Iron-chromium-nickel-manganese alloy steel. Etching: Use fresh acid.
	Ferric chloride	Ferric chloride .... 5 g Hydrochloric acid .... 20 ml Water, distilled .... 100 ml	Reveals structure of stainless and austenitic nickel steels.
	Marble's reagent	Cupric sulfate .... 4 g Hydrochloric acid .... 20 ml Water, distilled .... 20 ml	Reveals structure of various stainless steels.
High-speed steels	Snyder-Graff	Hydrochloric acid .... 9 ml Nitric acid .... 9 ml Methyl alcohol .... 100 ml	Reveals grain size of quenched and tempered high-speed steels. Etching time: 15 secs. to 5 min.

Chart R-13. Etching Reagents for Microscopic Examination of Iron and Steel.

	Etching Procedures		
Reagents	**Composition**	**Procedure**	**Uses**
Solutions for Aluminum			
Sodium hydroxide	NaOH . . . . 1 g Acetic acid . . . . 20 ml	Swab 10 seconds.	General microscopic.
Tucker's etch	HF. . . . . 15 ml HCl . . . . 45 ml HNO_3 . . . . 15 ml H_2O . . . . 25 ml	Etch by immersion.	Macroscopic.
Solutions for Stainless Steel			
Nitric and acetic acids	HNO_3 . . . . 30 ml Acetic acid . . . . 20 ml	Apply by swabbing.	Stainless alloys and others high in nickel or cobalt.
Cupric sulphate	$CuSO_4$ . . . . 4 gms HCl. . . . . 20 ml H_2O . . . . 20 ml	Etch by immersion. 10p11.75	Structure of stainless steels.
Cupric chloride and hydrochloric acid	$CuCl_2$ . . . . 5 gms HCl. . . . . 100 ml Ethyl alcohol. . . . . 100 ml H_2O . . . . 100 ml	Use cold immersion or swabbing.	Austenitic and ferritic steels.
Solutions for Copper and Brass			
Ammonium hydroxide & ammonium persulphate	NH_4OH. . . . . 1 part H_2O . . . . 1 part $(NH_4)_2S_2O_8$ (2 1/2%). 2 parts	Immersion	Polish attack of copper and some alloys.
Chromic acid	Saturated aqueous solution (CrO_3)	Immersion or swabbing	Copper, brass, bronze, nickel, silver (plain etch).
Ferric chloride	$FeCl_3$ . . . . 5 parts HCl . . . . 10 parts H_2O . . . . 100 parts	Immersion or swabbing (etch lightly)	Copper, brass, bronze, aluminum bronze.
Solutions for Iron and Steel			
Macro Examination			
Nitric acid	HNO_3 . . . . 5 ml H_2O . . . . 95 ml	Immerse 30 to 60 seconds.	Shows structure of welds.
Ammonium persulphate	$(NH_4)_2S_2O_3$ . . . . 10 gms H_2O . . . . 90 ml	Surface should be rubbed with cotton during etching.	Brings out grain structure, recrystallization at welds.
Nital	HNO_3 . . . . 5 ml Ethyl alcohol. . . . . 95 ml	Etch 5 min. followed by 1 sec. in HCl (10%).	Shows cleanness, depth of hardening, carburized or decarburized surfaces, etc.
Micro Examination			
Picric acid (picral)	Picric acid . . . . 4 gms Ethyl or methyl alcohol (95%) . . . . 100 ml	Etching time a few seconds to a minute or more.	For all grades of carbon steels.

Chart R-14. Etching Procedures.

			Alloy Wire Weights and Measures			
Brown & Sharpe Gage No.	Decimal	Phos. Bronze Ft. per lb.	18% Nickel Ft. per lb.	Aluminum Ft. per lb.	Copper Ft. per lb.	Brass Ft. per lb.
4/0	0.4600	1.559	1.599	5.207	1.561	1.640
3/0	0.4096	1.966	2.016	6.567	1.968	1.854
2/0	0.3648	2.480	2.543	8.279	2.482	2.068
1/0	0.3249	3.127	3.206	10.44	3.130	2.608
1	0.2893	3.943	4.043	13.16	3.947	3.289
2	0.2576	4.972	5.098	16.60	4.977	4.147
3	0.2294	6.269	6.428	20.94	6.276	5.229
4	0.2430	7.906	8.106	26.40	7.914	6.594
5	0.1819	9.969	10.22	22.20	9.980	8.315
6	0.1620	12.57	12.89	41.98	12.58	10.49
7	0.1443	15.85	16.25	52.91	15.87	13.22
8	0.1285	19.99	20.49	66.73	20.01	16.67
9	0.1144	25.20	25.84	84.19	25.23	21.02
10	0.1019	31.78	32.59	106.1	31.82	26.51
11	0.09074	40.08	41.09	133.9	40.12	33.43
12	0.08081	50.53	51.82	168.8	50.59	42.15
13	0.07196	63.72	65.39	212.5	63.80	53.15
14	0.06408	80.35	82.34	268.2	80.44	66.88
15	0.05707	101.3	103.9	337.9	101.4	84.68
16	0.05082	127.8	131.0	426.9	127.9	106.6
17	0.04526	161.1	165.2	536.9	161.3	134.4
18	0.04030	203.2	208.3	678.4	203.4	169.7
19	0.03589	256.2	262.7	854.9	256.5	213.7
20	0.03196	323.	331.2	1076.	323.4	269.5
21	0.02846	407.3	417.7	1356.	407.8	339.8
22	0.02535	513.6	526.7	1721.	514.2	428.5
23	0.02257	647.7	664.1	2157.	648.4	540.2
24	0.02010	816.7	837.4	2727.	817.7	681.3
25	0.01790	1030.	1056.	3439.	1031.	859.
26	0.01594	1299.	1332.	4358.	1300.	1083.
27	0.0142	1638.	1679.	5464.	1639.	1366.
28	0.01264	2065.	2117.	6940.	2067.	1723.

Chart R-15. Alloy Wire Weights and Measures.

Brinell		Vickers or Firth Hardness No.	Rockwell		Scleroscope	Tensile Strength 1000 psi
Dia. in mm, 3000 kg. Load 10 mm Ball	Hardness No.		C 150 kg. Load 120° Diamond Cone	B 100 kg. Load 1/16 in. dia. Ball		
2.05	898					440
2.10	857					420
2.15	817					401
2.20	780	1150	70		106	384
2.25	745	1050	68		100	368
2.30	712	960	66		95	352
2.35	682	885	64		91	337
2.40	653	820	62		87	324
2.45	627	765	60		84	311
2.50	601	717	58		81	298
2.55	578	675	57		78	287
2.60	555	633	55	120	75	276
2.65	534	598	53	119	72	266
2.70	514	567	52	119	70	256
2.75	495	540	50	117	67	247
2.80	477	515	49	117	65	238
2.85	461	494	47	116	63	229
2.90	444	472	46	115	61	220
2.95	429	454	45	115	59	212
3.00	415	437	44	114	57	204
3.05	401	420	42	113	55	196
3.10	388	404	41	112	54	189
3.15	375	389	40	112	52	182
3.20	363	375	38	110	51	176
3.25	352	363	37	110	49	170
3.30	341	350	36	109	48	165
3.35	331	339	35	109	46	160
3.40	321	327	34	108	45	155
3.45	311	316	33	108	44	150
3.50	302	305	32	107	43	146
3.55	293	296	31	106	42	142
3.60	285	287	30	105	40	138
3.65	277	279	29	104	39	134
3.70	269	270	28	104	38	131
3.75	262	263	26	103	37	128
3.80	255	256	25	102	37	125
3.85	248	248	24	102	36	122
3.90	241	241	23	100	35	119
3.95	235	235	22	99	34	116
4.00	229	229	21	98	33	113
4.05	223	223	20	97	32	110
4.10	217	217	18	96	31	107
4.15	212	212	17	96	31	104
4.20	207	207	16	95	30	101
4.25	202	202	15	94	30	99
4.30	197	197	13	93	29	97
4.35	192	192	12	92	28	95
4.40	187	187	10	91	28	93
4.45	183	183	9	90	27	91
4.50	179	179	8	89	27	89
4.55	174	174	7	88	26	87
4.60	170	170	6	87	26	85

Chart R-16. Hardness Conversion Table.

(Continued)

Hardness Conversion Table						
Brinell		Vickers or Firth Hardness No.	Rockwell		Scleroscope	Tensile Strength 1000 psi
Dia. in mm, 3000 kg. Load 10 mm Ball	Hardness No.		C 150 kg. Load 120° Diamond Cone	B 100 kg. Load 1/16 in. dia. Ball		
4.65	166	166	4	86	25	83
4.70	163	163	3	85	25	82
4.75	159	159	2	84	24	80
4.80	156	156	1	83	24	78
4.85	153	153		82	23	76
4.90	149	149		81	23	75
4.95	146	146		80	22	74
5.00	143	143		79	22	72
5.05	140	140		78	21	71
5.10	137	137		77	21	70
5.15	134	134		76	21	68
5.20	131	131		74	20	66
5.25	128	128		73	20	65
5.30	126	126		72		64
5.35	124	124		71		63
5.40	121	121		70		62
5.45	118	118		69		61
5.50	116	116		68		60
5.55	114	114		67		59
5.60	112	112		66		58
5.65	109	109		65		56
5.70	107	107		64		56
5.75	105	105		62		54
5.80	103	103		61		53
5.85	101	101		60		52
5.90	99	99		59		51
5.95	97	97		57		50
6.00	95	95		56		49

Chart R-16. *(Continued)*

Elements	Symbol	Density (Specific Gravity)	Weight per cu. ft.	Specific Heat	Melting Point	
					°C	°F
Aluminum	Al	2.7	166.7	0.212	658.7	1217.7
Antimony	Sb	6.69	418.3	0.049	630	1166
Armco iron	. . .	7.9	490.0	0.115	1535	2795
Carbon	C	2.34	219.1	0.113	3600	6512
Chromium	Cr	6.92	431.9	0.104	1615	3034
Columbium	Cb	7.06	452.54	. . .	1700	3124
Copper	Cu	8.89	555.6	0.092	1083	1981.4
Gold	Au	19.33	1205.0	0.032	1063	1946
Hydrogen	H	0.070*	0.00533	. . .	−259	−434.2
Iridium	Ir	22.42	1400.0	0.032	2300	4172
Iron	Fe	7.865	490.9	0.115	1530	2786
Lead	Pb	11.37	708.5	0.030	327	621
Manganese	Mn	7.4	463.2	0.111	1260	2300
Mercury	Hg	13.55	848.84	0.033	−38.7	−37.6
Nickel	Ni	8.80	555.6	0.109	1452	2645.6
Nitrogen	N	0.97*	0.063	. . .	−210	−346
Oxygen	O	1.10*	0.0866	. . .	−218	−360
Phosphorus	P	1.83	146.1	0.19	44	111.2
Platinum	Pt	21.45	1336.0	0.032	1755	3191
Potassium	K	0.87	54.3	0.170	62.3	144.1
Silicon	Si	2.49	131.1	0.175	1420	2588
Silver	Ag	10.5	655.5	0.055	960.5	1761
Sodium	Na	0.971	60.6	0.253	97.5	207.5
Sulfur	S	1.95	128.0	0.173	119.2	246
Tin	Sn	7.30	455.7	0.054	231.9	449.5
Titanium	Ti	5.3	218.5	0.010	1795	3263
Tungsten	W	17.5	1186.0	0.034	3000	5432
Uranium	U	18.7	1167.0	0.028		
Vanadium	V	6.0	343.3	0.115	1720	3128
Zinc	Zn	7.19	443.2	0.093	419	786.2
Bronze (90% Cu 10% Sn)	. . .	8.78	548.0	. . .	850–1000	1562–1832
Brass (90% Cu 10% Zn)	. . .	8.60	540.0	. . .	1020–1030	1868–1886
Brass (70% Cu 30% Zn)	. . .	8.44	527.0	. . .	900–940	1652–1724
Cast pig iron	. . .	7.1	443.2	. . .	1100–1250	2012–2282
Open-hearth steel	. . .	7.8	486.9	. . .	1350–1530	2462–2786
Wrought-iron bars	. . .	7.8	486.9	. . .	1530	2786

Properties of Elements and Metal Compositions

*Density compared with air.

Chart R-17. Properties of Elements and Metal Compositions.

Weights and Expansion Properties of Metals				
Metal	Weight per ft³ (lbs.)	Weight per m³ (kg)	Expansion per °F rise in temperature (0.0001 in.)	Expansion per °C rise in temperature (0.0001 mm)
Aluminum	165	2643	1.360	62.18
Brass	520	8330	1.052	48.10
Bronze	555	8890	0.986	45.08
Copper	555	8890	0.887	40.55
Gold	1200	19222	0.786	35.94
Iron (Cast)	460	7369	0.556	25.42
Lead	710	11373	1.571	71.83
Nickel	550	8810	0.695	31.78
Platinum	1350	21625	0.479	21.90
Silver	655	10492	1.079	49.33
Steel	490	7849	0.689	31.50

Chart R-18. Weights and Expansion Properties of Metals.

Safe Limits for Welding Fumes							
Material	Gases		Million Parts	Material	Gases		Million Parts
	ppm	mg/m³	per cu. ft.		ppm	mg/m³	per cu. ft.
Acetylene	1000			Titanium oxide		15.0	
Beryllium		0.002		Zinc oxide fumes		5.0	
Cadmium oxide fumes		0.1		Silica, crystalline			2.5
				Silica, amorphous			20.0
Carbon dioxide	5000			Silicates:			
Copper fumes		0.1		Asbestos			5.0
Iron oxide fumes		10.0		Portland cement			50.0
Lead		0.2		Graphite			15.0
Manganese		5.0		Nuisance dust			50.0
Nitrogen dioxide	5.0						
Oil mist		5.0					

Chart R-19. Safe Limits for Welding Fumes.

General Metric/U.S. Conventional Conversions			
Property	To convert from	To	Multiply by
Acceleration (angular)	revolution per minute squared	rad/s²	$1.745\ 329 \times 10^{-3}$
Acceleration (linear)	in/min²	m/s²	$7.055\ 556 \times 10^{-6}$
	ft/min²	m/s²	$8.466\ 667 \times 10^{-5}$
	in/min²	mm/s²mm/s²	$7.055\ 556 \times 10^{-3}$
	ft/min²	m/s²	$8.466\ 667 \times 10^{-2}$
	ft/s²		$3.048\ 000 \times 10^{-1}$
Angle, plane	deg	rad	$1.745\ 329 \times 10^{-2}$
	minute	rad	$2.908\ 882 \times 10^{-4}$
	second	rad	$4.848\ 137 \times 10^{-6}$
Area	in²	m²	$6.451\ 600 \times 10^{-4}$
	ft²	m²	$9.290\ 304 \times 10^{-2}$
	yd²	m²	$8.361\ 274 \times 10^{-1}$
	in²	mm²	$6.451\ 600 \times 10^{2}$
	ft²	mm²	$9.290\ 304 \times 10^{4}$
	acre (U.S. Survey)	m²	$4.046\ 873 \times 10^{3}$

Chart R-20. General Metric/U.S. Conventional Conversions.

(Continued)

Property	To convert from	To	Multiply by
Density	pound mass per cubic inch pound mass per cubic foot	kg/m^3 kg/m^3	$2.767\ 990 \times 10^4$ $1.601\ 846 \times 10$
Energy, work, heat, and impact energy	foot pound force foot poundal Btu* calorie* watt hour	J J J J J	$1.355\ 818$ $4.214\ 011 \times 10^{-2}$ $1.054\ 350 \times 10^3$ $4.184\ 000$ $3.600\ 000 \times 10^3$
Force	kilogram-force pound-force	N N	$9.806\ 650$ $4.448\ 222$
Impact strength	(see Energy)		
Length	in ft yd rod (U.S. Survey) mile (U.S. Survey)	m m m m km	$2.540\ 000 \times 10^{-2}$ $3.048\ 000 \times 10^{-1}$ $9.144\ 000 \times 10^{-1}$ $5.029\ 210$ $1.609\ 347$
Mass	pound mass (avdp) metric ton ton (short, 2000 lb/m) slug	kg kg kg kg	$4.535\ 924 \times 10^{-1}$ $1.000\ 000 \times 10^3$ $9.071\ 847 \times 10^2$ $1.459\ 390 \times 10$
Power	horsepower (550 ft lbs/s) horsepower (electric) Btu/min* calorie per minute* foot pound-force per minute	W W W W W	$7.456\ 999 \times 10^2$ $7.460\ 000 \times 10^2$ $1.757\ 250 \times 10$ $6.973\ 333 \times 10^{-2}$ $2.259\ 697 \times 10^{-2}$
Pressure	pound force per square inch bar atmosphere kip/in^2	kPa kPa kPa kPa	$6.894\ 757$ $1.000\ 000 \times 10^2$ $1.013\ 250 \times 10^2$ $6.894\ 757 \times 10^3$
Temperature	degree Celsius, $t°C$ degree Fahrenheit, $t°F$ degree Rankine, $t°R$ degree Fahrenheit, $t°F$ kelvin, t_K	K K	$t_K = t°C + 273.15$ $t_K = (t°F + 459.67)/1.8$ $t_K = t°R/1.8$ $t°C = (t°F - 32)/1.8$ $t°C = t_K - 273.15$
Tensile strength (stress)	ksi	MPa	$6.894\ 757$
Torque	inch pound force foot pound force	N·m N·m	$1.129\ 848 \times 10^{-1}$ $1.355\ 818$
Velocity (angular)	revolution per minute degree per minute revolution per minute	rad/s rad/s deg/min	$1.047\ 198 \times 10^{-1}$ $2.908\ 882 \times 10^{-4}$ $3.600\ 000 \times 10^2$
Velocity (linear)	in/min ft/min in/min ft/min mile/hour	m/s m/s mm/s mm/s km/h	$4.233\ 333 \times 10^{-4}$ $5.080\ 000 \times 10^{-3}$ $4.233\ 333 \times 10^{-1}$ $5.080\ 000$ $1.609\ 344$
Volume	in^3 ft^3 yd^3 in^3 ft^3 in^3 ft^3 gallon	m^3 m^3 m^3 mm^3 mm^3 L L L	$1.638\ 706 \times 10^{-5}$ $2.831\ 685 \times 10^{-2}$ $7.645\ 549 \times 10^{-1}$ $1.638\ 706 \times 10^4$ $2.831\ 685 \times 10^7$ $1.638\ 706 \times 10^{-2}$ $2.831\ 685 \times 10$ $3.785\ 412$

General Metric/U.S. Conventional

*Thermochemical

Chart R-20. *(Continued)*

Metric Units for Welding		
Property	**Unit**	**Symbol**
Area dimensions	Square millimeter	mm^2
Current density	Ampere per square millimeter	A/mm^2
Deposition rate	Kilogram per hour	kg/h
Electrical resistivity	Ohm meter	$\Omega \cdot m$
Electrode force (upset, squeeze, hold)	Newton	N
Flow rate (gas and liquid)	Liter per minute	L/min
Fracture toughness	Meganewton meter$^{-3/2}$	$MN \cdot m^{-3/2}$
Impact strength	Joule	$J = N \cdot m$
Linear dimensions	Millimeter	mm
Power density	Watt per square meter	W/m^2
Pressure (gas and liquid)	Kilopascal	$kPa = 1000\ N/m^2$
Tensile strength	Megapascal	$MPa = 1000\ 000\ N/m^2$
Thermal conductivity	Watt per meter Kelvin	$W/(m \cdot K)$
Travel speed	Millimeter per second	mm/s
Volume dimensions	Cubic millimeter	mm^3
Electrode feed rate	Millimeter per second	mm/s

Chart R-21. Metric Units for Welding.

Converting Measurements for Common Welding Properties			
Property	**To convert from**	**To**	**Multiply by**
Area dimensions (mm²)*	in^2 mm^2	mm^2 in^2	$6.451\ 600 \times 10^2$ $1.550\ 003 \times 10^{-3}$
Current density (A/mm²)	A/in^2 a/mm^2	A/mm^2 A/in^2	$1.550\ 003 \times 10^{-3}$ $6.451\ 600 \times 10^2$
Deposition rate** (kg/h)	lb/h kg/h	kg/h lb/h	0.045** 2.2**
Electrical resistivity (Ω·m)	Ω·cm Ω·m	Ω·m Ω·cm	$1.000\ 000 \times 10^{-2}$ $1.000\ 000 \times 10^2$
Electrode force (N)	pound-force kilogram-force N	N N ibf	4.448 222 9.806 650 $2.248\ 089 \times 10^{-1}$
Flow rate (L/min)	ft^3/h gallon per hour gallon per minute cm^3/min L/min cm^3/min	L/min L/min L/min L/min ft^3/min ft^3/min	$4.719\ 475 \times 10^{-1}$ $6.309\ 020 \times 10^{-2}$ 3.785 412 $1.000\ 000 \times 10^{-3}$ 2.118 880 $2.118\ 880 \times 10^{-3}$
Fracture toughness (MN·m⁻³/²)	$ksi·in^{1/2}$ $MN·m^{-3/2}$	$MN·m^{-3/2}$ $ksi·in^{1/2}$	1.098 855 0.910 038
Heat input (J/m)	J/in J/m	J/m J/in	$3.937\ 008 \times 10$ $2.540\ 000 \times 10^{-2}$
Impact energy	foot pound force	J	1.355 818
Linear measurements (mm)	in ft mm mm	mm mm in ft	$2.540\ 000 \times 10$ $3.048\ 000 \times 10^2$ $3.937\ 008 \times 10^{-2}$ $3.280\ 840 \times 10^{-3}$
Power density (W/m²)	W/in^2 W/m^2	W/m^2 W/in^2	$1.550\ 003 \times 10^3$ $6.451\ 600 \times 10^{-4}$
Pressure (gas and liquid) (kPa)	psi lb/ft^2 N/mm^2 kPa kPa kPa torr (mm Hg at 0°C) micron (μm Hg at 0°C) kPa kPa	Pa Pa Pa psi il/ft^2 N/mm^2 kPa kPa torr micron	$6.894\ 757 \times 10^3$ $4.788\ 026 \times 10$ $1.000\ 000 \times 10^6$ $1.450\ 377 \times 10^{-1}$ $2.088\ 543 \times 10$ $1.000\ 000 \times 10^{-3}$ $1.333\ 22\ \times 10^{-1}$ $1.333\ 22\ \times 10^{-4}$ $7.500\ 64\ \times 10$ $7.500\ 64\ \times 10^3$
Tensile strength (MPa)	psi lb/ft^2 N/mm^2 MPa MPa MPa	kPa kPa MPa psi lb/ft^2 N/mm^2	6.894 757 $4.788\ 026 \times 10^{-2}$ 1.000 000 $1.450\ 377 \times 10^2$ $2.088\ 543 \times 10^4$ 1.000 000
Thermal conductivity (W[m·K])	cal/(cmÃs·°c)	W/(m·K)	$4.184\ 000 \times 10^2$
Travel speed, electrode feed speed (mm/s)	in/min mm/s	mm/s in/min	$4.233\ 333 \times 10^{-2}$ 2.362 205

*Preferred units are given in parentheses.
**Approximate conversion.

Chart R-22. Converting Measurements for Common Welding Properties.

Glossary of Welding Terms

A

Abrasive pads: Commercial cleaning pad used to remove oxide films from weld joint area prior to welding.

Acetone: Colorless, volatile, water-soluble flammable liquid used to remove grease and oils from weld joint prior to welding.

Afterflow time: Period of time after welding that shielding gas flows through the welding gun to shield the electrode from contamination.

Age: Metallurgical term for the natural hardening sequence used to change the mechanical properties of a material.

Air carbon arc cutting (AAC): Arc cutting process in which metals to be cut are melted by heat of carbon arc. Molten metal is removed by a blast of compressed air.

Alkaline cleaner: A mixture of sodium hydroxide and water used for removing oil and oxides from metal. Commonly used on aluminum and magnesium for preweld cleaning.

Alloy: A material formed by the combination of two or more metallic elements.

Alternating current: Electrical current is the flow of electrons through a conductor. In alternating current the electrons flow in one direction, stop, and reverse the flow (alternate directions) for each complete cycle. In the United States, alternating current flows at 60 cycles per second. Therefore, the current stops and starts 120 times a second.

American Iron and Steel Institute (AISI): Industry association of iron and steel producers. It provides statistics on steel production and use, and publishes steel products manuals.

American Society for Testing Materials (ASTM): Organization that makes specifications for iron and steel products for industry.

American Welding Society (AWS): Nonprofit technical society organized and founded for the purpose of advancing the art and science of welding. The AWS publishes codes and standards concerning all phases of welding, and a magazine, *The Welding Journal.*

Ammeter: Instrument for measuring electrical current in amperes.

Amperage: The measurement of the amount of electricity flowing past a given point in a conductor per second. Current is another name for amperage.

Annealing: A process of softening metal by heating and slow cooling.

Anode: Positive terminal or pole of a circuit.

Arc blow: Deflection of intended arc pattern by magnetic fields. Also called "magnetic arc blow."

Arc gap: The physical gap between the end of the electrode and the base metal. The gap creates heat due to resistance of current flow and the arc rays.

Arc outages: Loss of arc during welding.

Arc voltage: Measurement of electrical potential (pressure or voltage) across the arc gap between the end of the electrode and the workpiece during the welding process.

Argon: Inert gas used to shield electrode and weld pool. It is heavier than air, colorless, and tasteless.

Arrow side: The area under the AWS welding symbol's horizontal reference line. A weld symbol placed there indicates that the weld is to be made on the side of the weld touched by the arrow.

Artificial age: Metallurgical term describing a heating operation in the hardening sequence used to change mechanical properties of metal.

Asphyxiation: Loss of consciousness due to a lack of oxygen; can be fatal.

Atmosphere: Envelope of gases in which the welding arc is enclosed. May also be called an inert atmosphere.

Austenite: Nonmagnetic stainless steel that cannot be hardened by heat treatment.

Austenitic stainless steel: Alloy of iron that contains at least eleven percent chromium with varying amounts of nickel. The grain structure is always austenitic (nonmagnetic).

Automatic welding: Welding that is mechanically moved while controls govern the speed and direction of the weld process.

AWS Welding symbol: See **Welding symbol.**

B

Backflow check valve: A one-way valve in a gas line, designed to prevent reverse flow that would result in improper mixing of gases.

Background current: Lowest amount of current (amperage) used to maintain the arc during a pulsing sequence.

Backhand welding: Welding with the gun pointed back toward the molten pool; also called "pull welding."

Backing bar: Tool or fixture attached to the root of weld joint. Tool may or may not control the shape of the penetrating metal.

Beryllium: Hard, light metallic element used in copper alloys for better fatigue endurance.

Bevel: Angular type of edge preparation.

Birdnest: A tangle of welding wire, usually caused by incorrect adjustment of the wire reel brake control.

Bore: Inside diameter of hole, tube, or hollow object.

Brass: Various metal alloys consisting mainly of copper and zinc.

Bridging fillet weld: A weld made when the parts to be welded are not in close contact.

Bright anneal: Process of annealing (softening) metal. Usually carried out in a controlled furnace atmosphere so surface oxidization is reduced to a minimum. Surface remains relatively bright.

Bright metal: Material preparation where the surface has been ground or machined to a bright surface to remove scale or oxides.

Brittle weld: Hard weld with little or no ductility.

Bronze: Various metal alloys consisting mainly of copper and tin.

Burnback control: Timer on a welding machine that prevents the wire from sticking in the weld pool. It sets the time that the arc power is on after the stop switch is released to burn back the wire from the molten weld pool.

Burn through: Weld that has melted through, resulting in a hole and excessive penetration.

Buttering: Form of surfacing where one or more layers of metal are placed on the weld groove face. Material then becomes transition weld deposit for final weld joint.

C

Cadmium: White ductile metallic element used for plating material to prevent corrosion. The fumes are toxic.

Carbide precipitation: Movement of chromium from grains into grain boundaries in the chrome-nickel stainless steels.

Carbides: Compound of carbon with one or more metal elements.

Carbon dioxide: A compound of carbon and oxygen used as a primary shielding gas or part of a mixed shielding gas combination. Symbol is CO_2.

Cast: Term referring to amount of curvature in coil of electrode wire.

Cathode: Negative terminal or pole of electrical circuit.

Certificate of Conformance: A statement that the filler material meets all of the requirements of the material specification.

Certified Chemical Analysis: Report of chemical analysis on particular heat, lot, or section of material.

Charpy impact test: Test used to determine resistance to failure caused at a notch at low temperatures.

Chemical composition: Makeup of a material, expressed in chemical percentages.

Chromium-molybdenum steels: Alloys used in applications requiring high strength. Special procedures are required when welding these materials to yield the desired mechanical properties.

Cladding: Layer of material applied to a surface to improve corrosion resistance.

Cold laps: Area of weld that has not fused with the base material.

Cold shut: Lack of fusion, usually occurring between passes in a multipass weld.

Cold working: Increasing the tensile properties of material by working the material without heat.

Concave weld crown: Weld crown that is curved inward.

Conduit: Metallic sheath or tube through which the filler wire is moved from wire feeder to weld area.

Constant current (CC): Term used for arc welding machines that produce a nearly constant current even though the arc gap voltage may vary.

Constant potential (CP): Another term for "constant voltage."

Constant voltage (CV): Term used for arc welding machines that produce a nearly constant voltage even when the current changes.

Contact tip: Part in a gun that conducts electrical current from power supply to consumable welding wire.

Contactor: Electric switch in power supply. Activating the contactor transfers welding power from the welding machine to the welding gun.

Contamination: Term used to indicate a dirty workpiece, impure shielding gas, improper gas shielding, or other problem that could result in weld porosity or other defects.

Contour: Shape of the weld bead or pass.

Convex weld crown: Weld crown that is curved outward.

Copper: Malleable, ductile, metallic element.

Corrective action: Action to be taken to prevent weld defects.

Corrosion: Eating away of material by a corrosive medium.

Corrosion resistance: Properties of a metal to resist chemical or electrochemical interaction with its surroundings, thus preventing it from deterioration.

Cracking: Operation of slightly opening a high-pressure gas cylinder valve to blow
out dirt.

Crater: Depression at the end of a weld that has insufficient cross section.

Crater cracks: Cracking that occurs in a crater.

Crown dimension: The height of the weld bead.

Cryogenic temperatures: Very cold temperatures; usually considered to be the temperatures of liquefied gases.

Cubic feet per hour (cfh): Unit measurement of the amount of gas flow used in gas shielded arc welding operations.

Current: The amount of electricity flowing past a point in a conductor every second. Also called amperage.

D

Defects: Imperfections that can occur in the weld or in the filler wire.

Demurrage: Monetary charge applied to the user of gas cylinders beyond the agreed rental period.

Deoxidized filler material: Filler material that contains deoxidizers such as aluminum, zirconium, and titanium for welding steels.

Department of Transportation (DOT): Government organization responsible for establishing and maintaining rules for the safe handling of gases used in welding.

Destructive testing (DT): Series of tests by destruction to determine the quality of a weld.

Dewar flask: Specially constructed tank similar to a vacuum bottle for the storage of liquefied gases.

Dies: Tools that are used to form wire or metal to a specified shape or dimension.

Direct current: Flow of current (electrons) in only one direction, either to the workpiece or to the electrode.

Direct current electrode negative (DCEN): Direct current flowing from electrode to the work.

Direct current electrode positive (DCEP): Direct current flowing from the work to the electrode.

Direct current reverse polarity (DCRP): Preferred terminology is *direct current electrode positive.*

Direct current straight polarity (DCSP): Preferred terminology is *direct current electrode negative.*

Double weld: A weld made from both sides of the joint.

Downhill: Welding direction. Preferred term for *vertical-down* welding.

Drag angles: Gun angles for backhand welding.

Drive rollers: Specially designed rollers for various types and sizes of filler wire to be fed through a mechanized wire feeder.

Droopers: Name given to constant current welding machines, since the slant of the volt-ampere curve they produce is drooping.

Dross: Oxidized metal or impurities within the metal.

Ductility: Property of material to deform permanently, or to exhibit plasticity without breaking while under tension.

Duty cycle: See **Machine duty cycle.**

E

Electrical stickout: Term used by manufacturers of welding materials referred to as "ESO" in welding procedures.

Electrode: In GMAW, the wire that is melted as a result of the electric arc and becomes part of the weld.

Electrode extension: The distance that the filler wire extends, measured from the contact tip to the end of the electrode. See **Stickout.**

Electrons: Negatively charged particles.

Elongation: Permanent elastic extension which metal undergoes during tensile testing. Amount of extension is usually indicated by percentages of original gauge length.

Excessive penetration: Extension of the weld below the root.

Excessive spatter: Scattering of more than the permissible amount of metal droplets in the vicinity of the weld.

F

Feathered tack weld: Tack weld that has been tapered on both ends to assist in obtaining the proper penetration during a groove joint root pass weld.

Ferrite test: Test of austenitic stainless steel weld deposit to determine amount of ferrite.

Ferritic stainless steel: Group of iron-chromium and carbon alloys that are nonhardenable by heat treatment.

Ferromagnetic: Term for a material that can possess magnetization in absence of an external magnetic field.

Ferrous metals: Group of metals containing substantial amounts of iron.

Filler wire: Welding material made in the form of wire which is added to the molten weld pool.

Fillet weld: Weld of approximately triangular cross section, used to join two surfaces approximately at right angles to each other in a lap joint, T-joint, or corner joint.

Fillet weld leg: Leg lengths of largest isosceles right triangle which can be inscribed within fillet weld cross section.

Fillet weld throat: Distance measured from bottom part of weld to crown surface.

Flowmeter: Mechanical device to measure rate of flow for gases used in welding. Usually measurement is in cubic feet per hour (cfh).

Flux-cored arc welding (FCAW): An arc welding process that produces coalescence of metals by heating them with an arc between a continuous filler metal electrode and the work. Shielding is provided by a flux contained within the electrode.

Forehand welding: Gun is pointed away from the molten pool, also called "push welding."

Free machining steels: Steels have been modified for use as machining stock by adding special materials. These steels are not weldable.

Fuse: Electrical circuit device designed to fail at a predetermined current level to give protection from overcurrent.

Fusion: Melting together of filler metal and base metal or of base metal itself.

Fusion welding: Process that uses fusion to complete the weld.

G

Galvanizing: Process in which a coating of zinc metal is applied to a steel surface to prevent corrosion.

Gas-cooled gun: A gun that uses the flow of shielding gas as its cooling medium.

Gas envelope: Shape and pattern of shielding gas over the weld area.

Gas metal arc welding (GMAW): Welding method using a consumable, continuously fed wire electrode and a gas that shields and protects the molten weld pool from atmospheric contamination. Also called "MIG welding."

Gas nozzle: Assembly made from glass, metal, or ceramics that attaches to gun body to direct shielding gas flow over weld area.

Gas tungsten arc welding (GTAW): Welding method using a nonconsumable tungsten electrode and a gas that shields and protects the molten weld pool from atmospheric contamination. Also called "TIG welding."

Globular arc mode: Method of weld metal deposition in form of larger globules of metal that detach from the wire on an indefinite pattern.

Grit blasting: Process for cleaning or finishing metal by the use of an air blast that blows particles of an abrasive (small pieces of steel, sand, steel balls) against the workpiece.

Ground connection: Safety connection from a welding machine frame to earth. Often used for grounding an engine-driven welding machine where a cable is connected from a ground stud on the welding machine to a metal stake placed in the ground.

Guarantee: See **Warranty.**

Gun: Trigger or switch-operated mechanism held by operator, or mounted in machine, that transfers electrical current to electrode. Typically contains a gas nozzle to direct shielding gas around molten pool.

Gun angle: Distance measured from perpendicular between gun and work surface. Tilting gun from perpendicular affects penetration, bead formation, and final bead appearance.

Gun cable: Stranded copper alloy cable which provides welding current from power supply to welding gun.

H

Hard surfacing: Abrasion-resistant material applied to surface of softer material for protection from wear.

Heat-affected zone: The area of the base metal around the weld joint that has been changed, mechanically or chemically, by the heat of the welding operation.

Heat-resistant: Term for materials that resist oxidation at high temperatures.

Heat sink: Tooling applied adjacent to weld zone to absorb heat and to prevent heat flow into parent material. Also called "chill bars."

Helium: The second lightest element, used as an inert shielding gas to increase penetration of weld and to increase welding speed.

Helix: The maximum height at any point of one complete circle of filler wire from the spool, as it lies on a flat surface.

Hot-drawn: Term applied to filler wire that has been drawn through forming dies while at an elevated temperature.

Hot shortness: Term used to describe metal (such as aluminum) that exhibits weakness when hot.

Hot start: Programmed period of time at the start of the weld sequence. Uses higher welding current than normally used to achieve deeper penetration.

Hydraulic systems: Pressure systems that use a fluid, such as oil, to move or hold weld parts or tooling.

Hydrogen: The lightest known element, sometimes used in mixtures of argon for the welding of stainless steels to achieve a clean weld surface.

I

Icicles: Intermittent sections of weld drop-through extending below the normal contour of full penetration groove weld.

Inconel: Nickel alloy that contains substantial percentages of chromium and iron.

Inductance: The squeezing force on an electrode, caused by current flow, that separates drops of molten filler metal. Also called "pinch effect."

Inert gas: Gas which does not normally combine chemically with base metal or filler material.

Ingot: Large block of metal, usually cast in metal mold. Forms basic material for further processing.

Interference fit: A fit used with cylindrical parts, in which parts are held tightly together by friction.

Intermittent weld: A seam weld that is broken by recurring unwelded spaces. Preferred term for *stitch weld.*

Interpass temperature: In multiple pass weld, minimum or maximum temperature specified for the deposited metal before the next weld pass is started.

Interstate Commerce Commission (ICC): Federal agency that once controlled testing and approval of cylinders used in transporting various types of gases used in the welding industry. The U.S. Department of Transportation now has this responsibility.

Izod test: Test used to determine notch impact values. Often used in conjunction with Charpy impact test.

J

Joggle-type joints: Joints used in welding auto body panels. A flange is formed on the lower piece of metal, so that a smooth upper surface can be achieved after welding.

Joint: Area where two or more pieces are brought together in an assembly.

L

Lack of fusion: Fusion which is less than complete.

Lack of penetration: Joint penetration which is less than specified.

Linear porosity: Cavity-type discontinuities formed by gas entrapment along a line during solidification of liquid melt.

Liners: Materials used to decrease friction and wear from filler wire on the inside of gun cables.

Liquefied gas: Gas that has been changed into a liquid for ease in storage and handling.

Liters per minute (lpm): Metric measurement of the amount of shielding gas flow used in the welding operation.

Load voltage: The actual voltage after the arc is established.

Longitudinal: In a direction *along* the weld, as opposed to *transverse* (across the weld).

M

Machine duty cycle: Period of time established by the machine manufacturer for operation within the machine's design specification.

Macro test: Test of a weld or parent metal structure cross section with magnification or with the naked eye.

Magnesium: Light, ductile, silver-white metal with high strength-to-weight ratio.

Magnetic arc blow: See **Arc blow.**

Magnetic field: Area where the magnetic lines of force have been established. Magnetic fields may be used to deflect the arc for oscillation of the molten pool.

Magnetic particle inspection: Nondestructive inspection test using a magnetized part and finely divided iron particles to outline discontinuities in the parent metal or the weld.

Manifold: Pipe or cylinder with several inlet and/or outlet fittings designed so several single cylinders can be fitted together to supply multiple welding stations.

Manipulator: Positioning tool used to locate welding equipment at the weld area for longseam or circumferential welding.

Martensite: Structure obtained when steel is heated and cooled to achieve its maximum hardness.

Mechanical properties: Includes such areas as tensile strength, ductility, brittleness, elasticity, hardness, toughness, and malleability.

Micro cracks: Very small cracks within metal. Located and seen only with the aid of very high magnification.

Microswitch: Switch often used in foot or hand controls for initiating welding cycles. Switch operates by depressing lever a few thousandths of an inch.

Mismatch: Junction point of butt joint where the top or bottom edges are not even.

Mode: Term used to describe set-up of a welding machine for manual, remote, or automatic operation.

Module: Unit which contains all the required circuits for a specified sequence. Made as controls for various areas such as pulsers, slope, and spot welding.

Motor-generator: Welding power supply using an electrical motor to drive a welding generator.

Multipass weld: A weld that is completed by making several passes (beads) next to or on top of each other.

N

National Electrical Manufacturers Association (NEMA): Industrial association of manufacturers of electrical machinery. Publishes standards and industry statistics, including welding.

Nick-break test: Destructive test of the weld to determine internal quality.

Nitrogen: Colorless, odorless, gaseous element.

Nondestructive test: Test which does not require destruction of the part to determine quality.

Nonferrous: Any metal which does not contain iron.

Nontoxic: Not harmful or poisonous.

Notch sensitivity: Resistance of metal to notch failure when subjected to rapid loading or stress.

Notch toughness: Ability of metal to resist failure (cracking) at a notch during loading (stress).

Nozzle: Ceramic or metal tube that attaches to the welding torch to direct the shielding gas flow over the weld area.

Numerical control: Numerical program established on tape or disk which controls the parameters and sequences of the welding operation.

O

Open circuit voltage: Voltage between the output terminals of an operating welding machine when no current is flowing in the circuit.

Oscillate: To move the welding gun back and forth across weld joint.

Other side: The area above the AWS welding symbol's horizontal reference line. A weld symbol placed there indicates that the weld is to be made on the side of the weld opposite the one touched by the arrow.

Outgassing: Sequence of allowing hydrogen gas to rise through the molten weld metal to the surface. When welding aluminum with a pulser, the pulsing action should be rapid to accomplish this sequence.

Out-of-position welding: Welding that is performed in one of the standard positions, such as horizontal, vertical, and overhead.

Output rating: Output limits of power supply with regard to current, open circuit voltage, ranges, power factor, and duty cycle.

Overlap: Protrusion of weld metal beyond the toe, face, or root of the weld.

Overlay: Weld placed over the top of another metal for dimensional requirements or to add physical or mechanical properties.

Oxidation: Process of reaction with an oxidizing agent.

Oxide film: Film formed on base material as a result of exposure to oxidizing agents, atmosphere, chemicals, or heat.

Oxygen: An atmospheric gas, always mixed in small quantities with other shielding gases, that is used to attain specific arc patterns.

P

Parameters: See **Welding parameters.**

Parts per million (ppm): Numerical method of identifying the purity or contamination of a gas.

Passes: Single weld beads are often termed passes.

Penetrant inspection: Visual surface inspection completed with dye and developer. May also use fluorescent dye and ultraviolet light for observing the results.

Penetration: The depth a weld extends into the joint and/or the base metal exclusive of reinforcement.

Phosgene gas: Poisonous, colorless, very volatile, suffocating gas. Made when trichloroethylene vapor enters the arc and is heated.

Physical properties: Properties of metals relating to electrical, thermal conductivity, and expansion rates.

Pinch: The physical force exerted on welding wire by the drive rollers of the wire feeder.

Plasma: Gas that has been heated to a partially ionized condition enabling it to conduct an electrical current.

Plasma arc cutting: Arc cutting process which severs or cuts metal by melting a localized area with a constricted arc. Process removes the molten metal with a high velocity jet of hot, ionized gas.

Polarity: Direction of current flow. Current flow from the electrode to the workpiece is DCEN. Current flow from the workpiece to the electrode is DCEP.

Porosity: A condition in which holes are located in the completed weld pass or layer of weld metal. The holes result from gas which formed during the melting of the material and was trapped in the cooling metal.

Positioner: Mechanical device for holding workpiece for welding in the desired position. It may rotate the weldment with a controlled speed for circular welds.

Post flow: Period of time at the end of the weld cycle where shielding gas flows around electrode during cooling to prevent contamination.

Postheat: Heat which is applied at the end of the weld cycle to slow down the cooling rate to prevent cracking and to relieve stresses.

Pounds per square inch (psi): Measurement of pressure.

Power supply: Special machine designed to produce the type and amount of welding current to melt the base metal and welding electrode.

Precipitate: Movement of elements within metal from grain into the grain boundary.

Preheat: Heat that is applied to the weldment prior to starting the welding operation. The actual heating temperature will vary depending on the type, thickness, and condition of the material to be welded.

Preheat temperature: Specified temperature base metal must be heated to immediately before welding is started.

Primary voltage: Incoming voltage of alternating current supplied by utility company.

Procedure: Sequence of events to be accomplished to make a weldment. This may include, but is not limited to: assembly sequence, cleaning procedure, tooling, tack welding, preheating, types of welding current, and current levels.

Programmer: Electronic or mechanical sequencer used to start and finish various portions or all of the sequences required to complete a weld.

Puddle: Incorrect term for molten portion of weld. See **Weld pool.**

Pull feeders: Wire feeders, usually located in the welding gun, that pull the wire through the cable to the contact tip.

Pull welding: See **Backhand welding.**

Pulsed MIG (MIG-P): A modified spray arc transfer that does not produce spatter because the wire does not touch the weld pool. Applications best suited for pulsed MIG are those currently using short circuit transfer method for welding steel, 14 ga. (1.8 mm) and thicker.

Pulser: Electronic device for sequencing and controlling amounts of current and time periods of operation.

Pulsing: Sequencing and control of the amount of current, and the duration of the welding arc.

Purging: Operation to remove air (atmosphere) from welding area and replace it with an inert atmosphere.

Push feeders: Wire feeders, usually located in the welding machine, that push the wire through the cable and gun.

Push-pull system: A wire-feed system that uses both types of feeder, usually one in the welding machine and the other in the gun.

Push welding: See **Forehand welding.**

Q

Qualified procedure: Welding sequence that has met all testing requirements included in the fabrication specification.

Qualified welder: Person who has demonstrated ability to produce a weld within the requirements of the fabrication specification.

Qualified welding operator: Person who has demonstrated the ability to operate a welding machine with a qualified procedure to produce a weld within the requirements of the fabrication specification.

R

Radiation: The energy that a welding arc radiates (sends out) in the form of intense ultraviolet light, or arc rays, that can burn the human body if it is not adequately protected.

Radiographic inspection: Nondestructive testing method used to determine interior quality of weld.

Rated load: Load (welding current) that may be obtained from power supply within limits established by the machine duty cycle.

Rectification: As applied to alternating current arc, refers to the device which acts as a one-way valve to convert one-half of the ac cycle to current flowing the same as the other half cycle.

Recrystallized: Term describing the new grain structure that has been obtained by heating a cold-worked metal.

Red-heat temperature: Temperature point in some metals at which the material becomes brittle.

Reference arc voltage: Actual welding arc voltage established on the automatic welding control circuit. By reference to this voltage, the arc voltage sensing control circuit maintains desired arc voltage.

Reference line: The horizontal line of the AWS welding symbol. All information about the weld to be made is positioned above or below this line.

Regulator: Valve to reduce and control gas pressure from cylinder to appropriate pressure for torch, used to keep pressure constant.

Regulator/flowmeter: Device that combines the functions of the regulator and the flowmeter.

Reverse polarity: Electron flow is from workpiece to electrode.

Rheostat: Adjustable resistor that allows resistance to be changed without opening the circuit in which it is connected.

Rimmed steel: Steel not completely deoxidized during manufacture. Outer rim of ingot contains high quality steel used to make steel welding wires.

Root penetration: The penetration which extends beyond the root of the welding joint.

Routing: Removal of metal from a weld using a routing tool. The tool may have various shapes and forms. The cutters may be tool steel or tungsten carbide.

S

Safety cap: The metal cap used to cover and protect the valve at the top of high-pressure gas cylinders.

Seamer: Machine designed to hold material for the purpose of making longseam welds.

Seamwelder: Seamer machine with attached equipment for making longseam welds.

Semiautomatic welding: Welding that is manually moved by a welder controlling the gun while special equipment feeds electrode and/or shielding gas. The welding operator controls the sequences and adjusts various parameters as required.

Semi-killed steel: Steel which has been partially deoxidized during solidification in the ingot mold.

Sensitivity: Response of the automatic voltage-controlled welding gun to varying arc lengths.

Sensitization range: Temperature range in stainless steels where carbide precipitation may take place.

Set-up: The process of preparing the welding machine and workpiece for welding.

Shielding gas: A single gas or a multiple gas mixture used to shield molten metal from atmospheric contamination.

Shop-aid-tooling: Tooling in the way of clamps, screws, metal/forms, etc., that may be used to hold parts in alignment for welding.

Short-arc: Mode of deposition that short-circuits, burns off, and deposits melted metal from the electrode into the weld joint.

Short-circuiting arc: See **Short-arc.**

Silicon: Nonmetallic element used in steelmaking. If present in molten weld pool, it will rise to the surface. It may cause weld pool movement control problems, and should be removed from the surface of the completed weld before making another pass.

Silicon-controlled rectifier (SCR): Solid-state diodes used to change alternating current into direct current.

Slope: Term used to describe the slant of the volt-amperage curve and the operating characteristics of the power supply under load.

Sodium hydroxide: Caustic soda used in a five percent concentration with water for removal of heavy oxide films on aluminum.

Solenoids: Electrically operated valves used to start and stop the flow of shielding gas to the welding gun.

Soluble dam: Paper dam used to plug tubes and pipes for purging. Flushing the tubes after welding dissolves the dam.

Solution heat-treated: Material that has been heated to a predetermined temperature for a suitable length of time to allow some element in the material to enter into "solid solution." Alloy is then quickly cooled to hold the element in this solution.

Spatter: Small pieces of metal that have been ejected from molten weld pool and attached to base material outside the weld.

Spectrometer: A testing machine used to prepare a precise chemical analysis of a material, such as filler wire.

Spool brake control: Control that stops wire spool rotation at end of welding.

Spool gun: A variation of the standard gun. The gun body includes a small spool of filler material, an electrical motor, and drive rolls to feed the wire.

Spot welding: Controlled weld cycle that produces small circular welds to fasten two pieces of sheet metal together.

Spray-arc mode: Method of weld metal deposition in form of small droplets of metal melted off the electrode a short distance above the workpiece.

Spray-arc pulse: A mode of metal deposition that involves pulsing a drop of filler material from the end of the electrode at a controlled time in the welding cycle.

Stainless steel: Alloy of iron, containing at least 11% chromium and some nickel, that resists almost all forms of rusting and corrosion.

Steel: Alloy of iron and carbon with varying amounts of other alloying elements for specific mechanical properties.

Stickout: Length of unmelted electrode or wire extending beyond the contact tip of the gun. See **Electrode extension.**

Stitch weld: See **Intermittent weld.**

Straight polarity: Arrangement of direct current arc welding leads in which the work is positive and electrode is negative.

Strain-hardened material: Material which has been strained by stretching, pulling, or forming to produce a grain structure with higher mechanical properties.

Strength-to-weight ratio: Term used to define material strength in relationship to material weight.

Stress relief: The process of heating the weldment to a specific temperature for a period of time to relieve the stresses formed during the welding operation.

Stringer bead: Type of weld bead made without oscillation (side-to-side motion).

Suck-back: Nonstandard term for concave root surface in a full penetration groove weld.

Surfacing: Applying material to the surface of another material for protection from chemicals, heat, wear, rust, etc.

Swaging: Operation or process of changing the shape of material with mechanical tools such as hammers or dies.

Synergic programs: Programs established by the factory that can be modified by the user. They are capable of regulating changes in the welding parameters to obtain the best possible welding conditions.

T

Tack weld: Weld made to hold parts of a weldment in alignment until final weld is made.

Taps: Male and female connection devices designed to complete a welding circuit for various amounts of welding current.

Teflon®: Insulating material used to make gaskets and insulators for welding torches, as well as low-friction liners for cables through which welding wire is pushed or pulled. Teflon tape is also used on threaded inert gas connections to make an effective seal.

Temper designations: These define ferrous materials with various mechanical properties that have been made by a series of basic treatments. Usually, materials with decreased hardness and increased toughness.

Tensile test: Test to determine mechanical properties of metal. Test is done by placing specially designed piece of metal in a tension machine and applying a load until the part fails.

Terminal blocks: Blocks of copper or bronze for attaching welding power and ground cables.

Test welds: Welds that are made to confirm a new or altered welding procedure.

Thermal-cut material: Material which has been cut by a melting process, such as the oxyfuel gas, plasma-arc, or carbon arc methods.

Thermal overload protector: Heat-sensing device which will open the electrical circuit on welding machine when certain temperature is reached. Used to protect machine from damage due to excessive heat.

Thermal treatment: Treatment of metal by the use of heat.

Three-phase current: Type of alternating current used to operate most welding power supplies.

Timers: Mechanical or electronic devices used to sequence various welding machine and other equipment functions.

Titanium: Light, silvery, lustrous, very hard, corrosion-resistant metallic element.

Tolerance: Permissible variation of a characteristic, variable, or parameter.

Toluene: Colorless, water-soluble, flammable liquid with a benzene-like odor. It is used for removing oils, grease, and paint from material.

Tractor: Electrical-mechanical vehicle used to transport a welding gun along a weld joint.

Transverse: Extending *across* the weld joint, as opposed to *longitudinal* (along the weld).

Trichloroethylene: Colorless, poisonous liquid solvent generally used in vapor state to remove oils, grease, and paint from material.

Tungsten: Rare metallic element with very high melting point, approximately 3410°C (6170°F).

Turbulence: In reference to shielding gases, this means disturbance from a smooth flow—irregular changes in speed and direction of flow, allowing atmosphere to enter weld zone.

U

Ultrasonic inspection: Nondestructive test method using high-frequency sounds to determine interior quality of a weld.

Underbead cracking: Cracking that occurs in the base metal adjacent to weld, usually caused by hydrogen being absorbed into weld and base metal during welding operation.

Undercut: Groove melted into base metal next to the toe or root of the weld and left unfilled by weld metal.

Uphill: Welding direction. Preferred term for *vertical-up* welding.

V

Variable voltage power supply: Term given to constant current power supply. Varying the welding voltage (arc gap) changes the output current level of the power supply.

Vertical-down: See **Downhill.**

Vertical-up: See **Uphill.**

Visual inspection: Inspection of the material or weld conducted by visual means.

Voids: Holes or areas within the completed weld.

Volt-ampere curve: Operating characteristics of welding power supply established by the manufacturer's design.

W

Warranty: Period of time a manufacturer guarantees the product.

Wash bead: See **Weave bead.**

Water-cooled gun: System that uses water to cool the welding gun and power cable. These guns operate at a higher duty cycle than gas-cooled guns .

Weave bead: Bead that is made with an oscillating (side-to-side) motion to widen it.

Weld chemistry: Chemical content of the completed weld.

Weld cycle: Complete series of events involved in making of weld.

Weld joint: Junction of members or edges of mating parts to be joined.

Weld pool: Molten part of the weld during the welding sequence.

Weld root: Deepest part of the weld into the base metal.

Weld schedule: Form used to record welding parameters, welding variables, and other data, so the weld can be duplicated at a future date.

Weld symbols: Standardized representations used for types and positions of welds, weld finishes, and welding techniques. They are placed in specified areas of the AWS welding symbol. See **Welding symbol.**

Weld throat: The distance from the top of the weld (crown) to the bottom of the weld (root).

Weld zone: See **Heat-affected zone.**

Welder: Person who performs a manual or semi-automatic welding operation.

Welding arc voltage: See **Arc voltage**.

Welding grade carbon dioxide: Carbon dioxide that has had the moisture removed during processing.

Welding operator: Person who operates automatic welding equipment.

Welding parameters: The range of operating values specified in the Welding Procedure Specification. These include: amperage, voltage, travel speed, shielding gas type, and flow rate.

Welding Procedure Specification (WPS): A document that list all of the variables and procedures required to perform a specific weld.

Welding Procedure Qualification: The demonstration that welds made by a specific procedure can meet prescribed standards.

Welding Procedure Qualification Record (WPQR): A document providing the actual welding variables used to produce an acceptable test weld, and the results of the test conducted on the weld to qualify a Welding Procedure Specification.

Welding speed: Rate at which electrode or wire passes along joint.

Welding symbol: Graphic representation developed by the American Welding Society. It is placed on drawings to indicate type of weld joint, placement of weld, and type of weld to be made.

Welding torch: See **Gun.**

Whiskers: Pieces of weld wire that extend through the weld joint.

Wire cast: See **Cast.**

Wire feeder: Electromechanical device designed to feed spooled filler wire into the weld pool at a controlled rate of speed.

Wire speed: The rate, usually measured in inches per minute (ipm), at which the filler wire is fed through the welding gun.

Wrought material: Material made by processes other than casting.

Index